Disclaimer

The publisher of this book is by no way associated with the National Institute of Standards and Technology (NIST). The NIST did not publish this book. It was published by 50 page publications under the public domain license.

50 Page Publications.

Book Title: A Comparison of hp-adaptive Strategies for Elliptic Partial Differential Equations (long version)

Book Author: William F. Mitchell; Marjorie A. McClain

Book Abstract: The hp version of the finite element method (hp-FEM) combined with adaptive mesh refinement is a particularly efficient method for solving partial differential equations because it can achieve a convergence rate that is exponential in the number of degrees of freedom. hp-FEM allows for refinement in both the element size, h, and the polynomial degree, p. Like adaptive refinement for the h version of the finite element method, a posteriori error estimates can be used to determine where the mesh needs to be refined, but a single error estimate can not simultaneously determine whether it is better to do the refinement by h or by p. Several strategies for making this determination have been proposed over the years. In this paper we summarize these strategies and present the results of a numerical experiment to study the convergence properties of these strategies.

Citation: NIST Interagency/Internal Report (NISTIR) - 7824

Keyword: elliptic partial differential equations; finite elements; hp-adaptive strategy; hp-FEM

NISTIR 7824

A Comparison of hp-adaptive Strategies for Elliptic Partial Differential Equations (Long Version)

William F. Mitchell
Marjorie A. McClain

National Institute of
Standards and Technology
U.S. Department of Commerce

NISTIR 7824

A Comparison of hp-adaptive Strategies for Elliptic Partial Differential Equations (Long Version)

William F. Mitchell
Marjorie A. McClain
Information Technology Laboratory
Applied and Computational Sciences Division

October 2011

U.S. Department of Commerce
Rebecca M. Blank, Acting Secretary

National Institute of Standards and Technology
Patrick D. Gallagher, Under Secretary for Standards and Technology and Director

A Comparison of hp-Adaptive Strategies for Elliptic Partial Differential Equations (Long Version)

William F. Mitchell [*]
Marjorie A. McClain
Applied and Computational Mathematics Division
National Institute of Standards and Technology
Gaithersburg, MD 20899-8910

Abstract

The hp version of the finite element method (hp-FEM) combined with adaptive mesh refinement is a particularly efficient method for solving partial differential equations because it can achieve a convergence rate that is exponential in the number of degrees of freedom. hp-FEM allows for refinement in both the element size, h, and the polynomial degree, p. Like adaptive refinement for the h version of the finite element method, a posteriori error estimates can be used to determine where the mesh needs to be refined, but a single error estimate cannot simultaneously determine whether it is better to do the refinement by h or by p. Several strategies for making this determination have been proposed over the years. In this paper we summarize these strategies and present the results of a numerical experiment to study the convergence properties of these strategies.

Keywords: elliptic partial differential equations, finite elements, hp-adaptive strategy, hp-FEM

1 Introduction

The numerical solution of partial differential equations (PDEs) is the most compute-intensive part of a wide range of scientific and engineering applications. Consequently the development and application of faster and more accurate methods for solving partial differential equations has received much attention in the past fifty years. Many of the applications at the cutting edge of research are extraordinarily challenging. For these problems it is necessary to allocate computing resources in an optimal way in order to have any chance at solving the problem. Determining the best grid and approximation space on which to compute the solution is a central concern in this regard. Unfortunately, it is rarely possible to determine an optimal grid in advance. Thus, developing self-adaptive techniques which lead to optimal resource allocation is critical for future progress in many fields.

Self-adaptive methods have been studied for over 30 years now. They are often cast in the context of finite element methods, where the domain of the PDE is partitioned into a mesh consisting of a number of elements (in two dimensions, usually triangles or rectangles), and the approximate solution is a polynomial over each element. Most of the work has focused on h-adaptive methods. In these methods, the mesh size,

---

[*] Contribution of NIST, not subject to copyright.

h, is adapted locally by means of a local error estimator with the goal of placing the smallest elements in the areas where they will do the most good. In particular, elements that have a large error estimate get refined so that ultimately the error estimates, and presumably the error, are approximately equal over all elements. h-adaptive methods are quite well understood now, and are beginning to be used in practice.

Recently, the research community has begun to focus more attention on hp-adaptive methods. In these methods, one not only locally adapts the size of the mesh, but also the degree of the polynomials, p. The attraction of hp-adaptivity is that the error converges at an exponential rate in the number of degrees of freedom, as opposed to a polynomial rate for fixed p. Much of the theoretical work showing the advantages of hp-adaptive methods was done in the 1980's, but it wasn't until the 1990's that practical implementation began to be studied. The new complication is that the local error estimator is no longer sufficient to guide the adaptivity. It tells you which elements should be refined, but it does not indicate whether it is better to refine the element by h or by p. A method for making that determination is called an hp-adaptive strategy. A number of strategies have been proposed, but it is not clear which ones perform best under different situations, or even if any of the strategies are good enough to be used as a general purpose solver. In this paper we present an experimental comparison of several hp-adaptive strategies.

Any study of this type is necessarily limited in scope. The comparison will be restricted to steady-state linear elliptic partial differential equations on bounded domains in two dimensions with Dirichlet, natural or mixed boundary conditions. The standard Galerkin finite element method will be used with the space of continuous piecewise polynomial functions over triangles that are refined by the newest node bisection method.

The remainder of the paper is organized as follows. In Section 2 we define the equation to be solved, present the finite element method, and give some *a priori* error estimates. In Section 3 we give the details of the hp-adaptive finite element algorithm used in the experiments. Section 4 defines the hp-adaptive strategies to be compared. Section 5 contains the results of the experiments. Finally, we draw our conclusions in Section 6.

2 The Finite Element Method

We consider the elliptic partial differential equation

$$Lu = -\frac{\partial}{\partial x}\left(p(x,y)\frac{\partial u}{\partial x}\right) - \frac{\partial}{\partial y}\left(q(x,y)\frac{\partial u}{\partial y}\right) + r(x,y)u = f(x,y) \quad \text{in } \Omega \tag{1}$$

$$u = g_D(x,y) \quad \text{on } \partial\Omega_D \tag{2}$$

$$Bu = p(x,y)\frac{\partial u}{\partial x}\frac{\partial y}{\partial s} - q(x,y)\frac{\partial u}{\partial y}\frac{\partial x}{\partial s} + c(x,y)u = g_N(x,y) \quad \text{on } \partial\Omega_N \tag{3}$$

where Ω is a bounded, connected, polygonal, open region in R^2 with boundary $\partial\Omega = \partial\Omega_D \cup \partial\Omega_N$, $\partial\Omega_D \cap \partial\Omega_N = \emptyset$. Differentiation with respect to s is with respect to a counterclockwise parameterization of the boundary $(x(s),y(s))$ with $(dx/ds\, dy/ds) = 1$. If $c=0$ Equation 3 is the natural boundary condition. If, in addition, $p=q=1$ or $\partial\Omega_N$ consists of line segments that are parallel to the axes, Equation 3 is the Neumann boundary condition. We assume the data in Equations 1-3 satisfy the usual ellipticity and regularity assumptions. In one of the test problems, we extend the equation to a system of two equations containing a cross derivative term $\partial^2 u/\partial x \partial y$, and in another test problem we include first order derivative terms.

As usual, define the space L^2 by

$$L^2(\Omega) = \{v(x,y): \int_\Omega v^2 \, dxdy < \infty\}$$

with inner product

$$\langle u,v \rangle_2 = \int_\Omega uv \, dxdy$$

and norm

$$\|v\|_2^2 = \langle v,v \rangle_2.$$

We denote by $H^m(\Omega)$ the usual Sobolov spaces

$$H^m(\Omega) = \{v \in L^2(\Omega): D^\alpha v \in L^2(\Omega), |\alpha| \le m\}$$

where

$$D^\alpha v = \frac{\partial^{|\alpha|} v}{\partial^{\alpha_1} x \partial^{\alpha_2} y}, \alpha = (\alpha_1, \alpha_2), \alpha_i \in \mathbb{N}, |\alpha| = \alpha_1 + \alpha_2.$$

The Sobolov spaces have inner products

$$\langle u,v \rangle_{H^m(\Omega)} = \int_\Omega \sum_{|\alpha| \le m} D^\alpha u D^\alpha v \, dxdy$$

and norms

$$\|v\|_{H^m(\Omega)}^2 = \langle v,v \rangle_{H^m(\Omega)}.$$

We will also refer to the seminorm $|v|_{H^m(\Omega)}$ where the sum is over $|\alpha| = m$.

Let $H_0^m(\Omega) = \{v \in H^m(\Omega): v = 0 \text{ on } \partial\Omega_D\}$. Let $\tilde{u}_D$ be a lift function satisfying the Dirichlet boundary conditions in Equation 2 and define the affine space $\tilde{u}_D + H_0^1(\Omega) = \{\tilde{u}_D + v : v \in H_0^1(\Omega)\}$. Define the bilinear form

$$B(u,v) = \int_\Omega \left(p \frac{\partial u}{\partial x} \frac{\partial v}{\partial x} + q \frac{\partial u}{\partial y} \frac{\partial v}{\partial y} + ruv \right) dxdy + \int_{\partial\Omega_N} cuv \, ds$$

and the linear form

$$L(v) = \int_\Omega fv \, dxdy + \int_{\partial\Omega_N} g_N v \, ds$$

Then the variational form of the problem is to find the unique $u \in \tilde{u}_D + H_0^1(\Omega)$ that satisfies

$$B(u,v) = L(v) \quad \forall v \in H_0^1(\Omega).$$

The energy norm of $v \in H_0^1$ is defined by $\|v\|_{E(\Omega)}^2 = B(v,v)$.

The finite element space is defined by partitioning Ω into a grid (or mesh), G_{hp}, consisting of a set of N_T triangular elements, $\{T_i\}_{i=1}^{N_T}$ with $\bar\Omega = \cup_{i=1}^{N_T} \bar{T}_i$. If a vertex of a triangle is contained in the interior of an edge of another triangle, it is called a hanging node. We only consider compatible grids with no hanging nodes, i.e. $T_i \cap T_j, i \ne j$, is either empty, a common edge, or a common vertex. The diameter of the element is denoted h_i. With each element we associate an integer degree $p_i \ge 1$. The finite element space V_{hp} is the space of continuous piecewise polynomial functions on Ω such that over element T_i it is a polynomial of degree p_i. The degree of an edge is determined by applying either a minimum rule or a maximum rule over G_{hp} in which the edge is assigned the minimum or maximum of the degrees of the adjacent elements, respectively.

The finite element solution is the unique function $u_{hp} \in \tilde{u_D} + V_{hp}$ that satisfies

$$B(u_{hp}, v_{hp}) = L(v_{hp}) \quad \forall v_{hp} \in V_{hp}.$$

The error is defined by $e_{hp} = u - u_{hp}$.

The finite element solution is expressed as a linear combination of basis functions $\{\varphi_i\}_{i=1}^{N}$ that span $\tilde{u_D} + V_{hp}$,

$$u_{hp} = \sum_{i=1}^{N} \alpha_i \varphi_i(x,y)$$

For high order elements, there are a number of basis sets used in practice. A number of the hp strategies of Section 4 rely on the basis being a p-hierarchical basis in which the basis functions for a space of degree p are a subset of the basis functions for a space of degree p+1. In the results of Section 5 the p-hierarchical basis of Szabo and Babuška[33], which is based on Legendre polynomials, is used. Regardless of the choice of basis set, for an element of degree p_i with edge degrees $p_{i,j}$, $j=1,2,3$, there is one linear basis function associated with each vertex, $p_{i,j} - 1$ basis functions (one each of degree $2, 3 \ldots p_{i,j}$) associated with edge j, and $q - 2$ basis functions of degree q for $q = 3, 4 \ldots p_i$ (for a total of $(p_i - 1)(p_i - 2)/2$) whose support is the interior of the triangle.

The discrete form of the problem is a linear system of algebraic equations

$$Ax = b \tag{4}$$

where the matrix A is given by $A_{ij} = B(\varphi_i, \varphi_j)$ and the right hand side is given by $b_i = L(\varphi_i)$.

If h and p are uniform over the grid, $u \in H^m(\Omega)$, and the other usual assumptions are met, then the *a priori* error bound is [6,7]

$$\|e_{hp}\|_{H^1(\Omega)} \leq C \frac{h^\mu}{p^{m-1}} \|u\|_{H^m(\Omega)} \tag{5}$$

where $\mu = \min(p, m - 1)$ and C is a constant that is independent of h, p and u, but depends on m. The same references show that under certain conditions, such as Laplace's equation on a domain with reentrant corners, the exponent on p can be doubled to $-2(m - 1)$, i.e., the p-version of the finite element method converges twice as fast as the h-version.

With a suitably chosen hp mesh, and other typical assumptions, the error can be shown [14] to converge exponentially in the number of degrees of freedom,

$$\|e_{hp}\|_{H^1(\Omega)} \leq C_1 e^{-C_2 N^{1/3}} \tag{6}$$

for some C_1 and $C_2 > 0$ independent of N.

3 hp-Adaptive Refinement Algorithm

One basic form of an hp-adaptive algorithm is given in Figure 1. If the algorithm is run on a parallel computer, a load balancing step is performed either before or after the coarsening/refinement part of the algorithm.

There are a number of approaches to each of the steps of the algorithm. In particular:

```
begin with a very coarse grid
form and solve the linear system
repeat
   determine which elements to coarsen and whether to coarsen by h or p
   coarsen elements
   repeat
      determine which elements to refine and whether to refine by h or p
      refine elements
   until some criterion on amount of refinement is met
   form and solve the linear system
until some termination criterion is met
```

Figure 1: Basic form of an hp-adaptive algorithm.

- How is an element h-refined?
- How is an element p-refined?
- What error indicator is used to guide adaptive refinement?
- When is the program terminated?
- How is an element coarsened?
- How do you determine which elements should be coarsened?
- How do you determine which elements should be refined?
- How much refinement should occur before the linear system is formed and solved again?
- Should an element be refined by h or p?

Other considerations, that are beyond the scope of this paper, include how to create the initial grid, and how to solve the linear system.

Complete coverage of the possible answers to these questions is beyond the scope of this paper. We will focus on the choices used for the results given in Section 5, and in some cases briefly mention other possibilities or give a reference, but this is not intended to be exhaustive. Note that some of the hp strategies in Section 4 require a different choice, or even a modification of the basic algorithm. These exceptions will be noted in Section 4.

There are several ways to refine triangles [19]. In this paper, the newest node bisection method [20] is used. Briefly, a parent triangle is h-refined by connecting one of the vertices to the midpoint of the opposite side to form two new child triangles. The most recently created vertex is chosen as the vertex to use in this bisection. Triangles are always refined in pairs (except when the edge to be refined is on the boundary) to maintain compatibility of the grid. This may require first refining a neighbor triangle to create the second triangle of the pair. The h-refinement level, l_i, of a triangle T_i is one more than the h-refinement level of the parent, with level 0 assigned to the triangles of the initial coarse grid. p-refinement is fairly universally accepted as increasing the degree of the element by one, followed by enforcing either the minimum rule or maximum rule for the edges. We will use the minimum rule.

Adaptive refinement is guided by a local *a posteriori* error indicator computed for each element. There are several choices of error indicators; see for example [2,34]. For this paper, the error indicator for element T_i is given by solving a local Neumann residual problem:

$$L e_i = f - L u_{hp} \quad \text{in } T_i \quad (7)$$
$$e_i = 0 \quad \text{on } \partial T_i \cap \partial \Omega_D \quad (8)$$
$$B e_i = g_N - B u_{hp} \quad \text{on } \partial T_i \cap \partial \Omega_N \quad (9)$$
$$B e_i = -\frac{1}{2}\frac{\partial u_{hp}}{\partial n} \quad \text{on } (\partial T_i \setminus \partial \Omega_D) \setminus \partial \Omega_N \quad (10)$$

where L, B, f, g_N, $\partial \Omega_D$, and $\partial \Omega_N$ are defined in Equations 1-3, $\frac{\partial u_{hp}}{\partial n}$ is the jump in the outward normal derivative of u_{hp} across the element boundary, including the coefficients of the natural boundary conditions, and in Equation 10 B is modified by setting $c(x,y)=0$. If the degree of T_i is p_i, the approximate solution, $e_{i,hp}$ of Equations 7-10 is computed using the hierarchical bases of exact degree p_i+1. The error indicator for element T_i is then given by

$$\eta_i = \|e_{i,hp}\|_{E(T_i)}$$

A global energy norm error estimate is given by

$$\eta = \left(\sum_{i=1}^{N_T} \eta_i^2 \right)^{1/2}.$$

One criterion for program termination is that η be smaller than some prescribed error tolerance τ, or, to base it on the relative error rather than the absolute error, $\eta < \tau \|u_{hp}\|_{E(\Omega)}$. Other possibilities are to terminate when some quantity, such as number of elements, number of degrees of freedom, amount of memory, amount of computation time, etc., is reached, or combinations of criteria. In this paper, the primary termination criterion is a relative error tolerance, with number of degrees of freedom as a secondary criterion.

Coarsening of elements may be performed to reverse bad decisions about what refinements to perform, or to allow the grid to follow the behavior of the solution in a time dependent problem. Elements are h-coarsened by reversing the h-refinement, i.e., joining the child triangles back together to form the parent triangle. p-coarsening means decreasing the degree of the element by one, and enforcing the minimum or maximum rule for the edges. For steady state problems, one choice of which elements to coarsen is the empty set, i.e., don't perform coarsening. Other than that choice, the most common approach is to coarsen elements that have a sufficiently small error indicator, subject to any requirements for compatibility of the grid. In the numerical results of this paper, an element is coarsened if $\eta_i < \max_i \eta_i / 100$. The value 100 is arbitrary.

The elements that are refined are usually those that have a sufficiently large error indicator. Perhaps the most common approach is to refine those with an error indicator that is larger than some fraction, typically between 1/4 and 1/2, of the maximum error indicator. Another approach, which is used in this paper, is to refine those with $\eta_i > \tau \|u_{hp}\|_{E(\Omega)} / \sqrt{N_T}$. Note that if every element had $\eta_i = \tau \|u_{hp}\|_{E(\Omega)} / \sqrt{N_T}$ then $\eta / \|u_{hp}\|_{E(\Omega)} = \tau$, hence the $\sqrt{N_T}$ factor.

There are many ways to determine how much refinement to do before forming and solving the linear system. One could refine until the global error estimate has been reduced by some factor, such as 1/2 or 1/4, or one could refine until some quantity, e.g. number of elements or degrees of freedom, has been increased by some factor, such as 2 or 4. Both of these require that reasonable error indicators can be computed on the child elements. The approach taken in this paper is to perform the refine loop once. The downside of this approach is that it requires more passes through the outer loop, which means forming and solving the linear system more times. But for the purpose of this paper, which is to determine the convergence rate of various hp-adaptive strategies with respect to number of degrees of freedom, the excess computation time is not important.

The method for determining whether an element should be refined by h or by p is called an hp-adaptive strategy. Several strategies have been proposed over the years. Many of them will be described in the next section.

4 The hp-Adaptive Strategies

In this section, the hp-adaptive strategies that have been proposed in the literature are presented. In some cases, these strategies were developed in the context of 1D problems, rectangular elements, or other settings that are not fully compatible with the context of this paper. In those cases, the strategy is appropriately modified for 2D elliptic PDEs and newest node bisection of triangles.

4.1 Use of apriori Knowledge of Solution Regularity

It is well known that for smooth solutions p-refinement will produce an exponential rate of convergence, but near singularities p-refinement is less effective than h-refinement. This is a consequence of the *apriori* error bound in Equation 5. For this reason, many of the hp strategies use h-refinement in areas where the solution is irregular (i.e., locally fails to be in H^m for some finite m, also called nonsmooth) or nearly irregular, and p-refinement elsewhere. The simplest strategy is to use any *apriori* knowledge about irregularities. For example, it is known that linear elliptic PDEs with smooth coefficients and piecewise analytic boundary data will have point singularities only near reentrant corners of the boundary and where boundary conditions change [4]. Another example would be a situation where one knows the approximate location of a shock in the interior of the domain.

An hp-adaptive strategy of this type was presented by Ainsworth and Senior [4]. In this approach they simply flag vertices in the initial mesh as being possible trouble spots. During refinement an element is refined by h if it contains a vertex that is so flagged, and by p otherwise. We will refer to this strategy by the name APRIORI.

We extend this strategy to allow more general regions of irregularity, and to provide the strength of the irregularity. The user provides a function that, given an element T_i as input, returns a regularity value for that element. For true singularities, it would ideally return the maximum value of m such that $u \in H^m(T_i)$. But it can also indicate that a triangle intersects an area that is considered to be nearly irregular, like a boundary layer or sharp wavefront. Based on the definition of μ in Equation 5, if the current degree of the triangle is p_i and the returned regularity value is m_i, we do p-refinement if $p_i \leq m_i - 1$ and h-refinement otherwise. The same approach is used in all the following strategies that estimate the local regularity m_i.

4.2 Estimate Regularity Using Smaller p Estimates

Süli, Houston and Schwab [32] presented a strategy based on Equation 5 and an estimate of the convergence rate in p using error estimates based on $p_i - 2$ and $p_i - 1$. We will refer to this strategy as PRIOR2P. This requires $p_i \geq 3$, so we always use p-refinement in elements of degree 1 and 2.

Suppose the error estimate in Equation 5 holds on individual elements and that the inequality is an approximate equality. Let η_{i,p_i-2} and η_{i,p_i-1} be *a posteriori* error estimates for partial approximate solutions over triangle T_i using the bases up to degree $p_i - 2$ and $p_i - 1$, respectively. Then

$$\frac{\eta_{i,p_i-1}}{\eta_{i,p_i-2}} \approx \left(\frac{p_i - 1}{p_i - 2}\right)^{-(m_i - 1)}$$

and thus the regularity is estimated by

$$m_i \approx 1 - \frac{\log(\eta_{i,p_i-1}/\eta_{i,p_i-2})}{\log((p_i - 1)/(p_i - 2))}$$

Use p-refinement if $p_i \leq m_i - 1$ and h-refinement otherwise.

Thanks to the p-hierarchical basis, the computation of the error estimates is very inexpensive. For $1 \leq j < p_i$,

$$u_{hp}|_{T_i} = \sum_{\substack{\text{supp}(\varphi_k) \cap T_i = \emptyset}} \alpha_k \varphi_k = \sum_{\substack{\text{supp}(\varphi_k) \cap T_i = \emptyset \\ \deg(\varphi_k) \leq p_i - j}} \alpha_k \varphi_k + \sum_{\substack{\text{supp}(\varphi_k) \cap T_i = \emptyset \\ \deg(\varphi_k) > p_i - j}} \alpha_k \varphi_k$$

where $\text{supp}(\varphi_k)$ is the support of φ_k and $\deg(\varphi_k)$ is the degree of φ_k. The last sum is the amount by which the solution changed when the degree of the element was increased from $p_i - j$ to p_i, and provides an estimate of the error in the partial approximate solution of degree $p_i - j$ given in the next to last sum. (Indeed, the local Neumann error estimator of Equations 7-10 approximates this quantity for the increase from degree p_i to $p_i + 1$.) Thus the error estimates are

$$\eta_{i,p_i-j} = \left\| \sum_{\substack{\text{supp}(\varphi_k) \cap T_i = \emptyset \\ \deg(\varphi_k) > p_i - j}} \alpha_k \varphi_k \right\|_{H^1(T_i)}$$

which only involves computing the norm of known quantities.

4.3 Type parameter

Gui and Babuška[13] presented an hp-adaptive strategy using what they call a type parameter, γ. This strategy is also used by Adjerid, Aiffa and Flaherty[1]. We will refer to this strategy as TYPEPARAM.

Given the error estimates η_{i,p_i} and η_{i,p_i-1}, define

$$R(T_i) = \begin{cases} \frac{\eta_{i,p_i}}{\eta_{i,p_i-1}} & \eta_{i,p_i-1} = 0 \\ 0 & \eta_{i,p_i-1} = 0 \end{cases}$$

By convention, $\eta_{i,0} = 0$, which forces p-refinement if $p_i = 1$.

R is used to assess the perceived solution smoothness. Given the type parameter, $0 \leq \gamma < \infty$, element T_i is said to be of h-type if $R(T_i) > \gamma$, and of p-type if $R(T_i) \leq \gamma$. If element T_i is selected for refinement, then refine it by h-refinement if it is of h-type and p-refinement if it is of p-type. Note that $\gamma = 0$ gives pure h-refinement and $\gamma = \infty$ gives pure p-refinement.

For the error estimates, we use the local Neumann error estimate of Equations 7-10 for η_{i,p_i}, and the η_{i,p_i-1} of Section 4.2. For the results of Section 5, we use $\gamma = 0.3$ if the solution has a singularity, and $\gamma = 0.6$ otherwise.[1]

---
[1] The value for this parameter, and the parameters of the other strategies, was determined by a preliminary experiment to determine a single value (or possibly two values dependent on singularness) that generally works best, using a subset of the test problems.

4.4 Estimate Regularity Using Larger p Estimates

Another approach that estimates the regularity is given by Ainsworth and Senior [3]. This strategy uses three error estimates based on spaces of degree p_i+1, p_i+2 and p_i+3, so we refer to it as NEXT3P.

The error estimate used to approximate the regularity is a variation on the local Neumann residual error estimate given by Equations 7-10 in which Equation 10 is replaced by

$$Be_i = g_i \text{ on} (\partial T_i \setminus \partial\Omega_D) \setminus \partial\Omega_N$$

where g_i is an approximation of B that satisfies an equilibrium condition. This is the equilibrated residual error estimator in [2].

The local problem is solved on element T_i three times using the spaces of degree p_i+q, $q=1,2,3$, to obtain error estimates $e_{i,q}$. In contrast to the local Neumann residual error estimate, the whole space over T_i is used, not just the p-hierarchical bases of degree greater than p_i. These approximations to the error converge to the true solution of the residual problem at the same rate the approximate solution converges to the true solution of Equations 1-3, i.e.

$$\|e_i - e_{i,q}\|_{E(T_i)} \approx C(p_i+q)^{-\alpha}$$

where C and α are positive constants that are independent of q but depend on T_i. Using the Galerkin orthogonality

$$\|e_i - e_{i,q}\|^2_{E(T_i)} = \|e_i\|^2_{E(T_i)} - \|e_{i,q}\|^2_{E(T_i)}$$

this can be rewritten

$$\|e_i\|^2_{E(T_i)} - \|e_{i,q}\|^2_{E(T_i)} \approx C^2(p_i+q)^{-2\alpha}.$$

We can compute $\|e_{i,q}\|^2_{E(T_i)}$ and p_i+q for $q=1,2,3$ from the approximate solutions, so the three constants $\|e_i\|_{E(T_i)}$, C and α can be approximated by fitting the data. Then, using the *a priori* error estimate in Equation 5, the approximation of the local regularity is $m_i=1+\alpha$. Use p-refinement if $p_i \leq m_i - 1$ and h-refinement otherwise.

4.5 Texas 3 Step

The Texas 3 Step strategy [8, 23, 24] first performs h-refinement to get an intermediate grid, and follows that with p-refinement to reduce the error to some given error tolerance, τ. We will refer to this strategy as T3S. Note that for this strategy the basic form of the hp-adaptive algorithm is different than that in Figure 1.

The first step is to create an initial mesh with uniform p and nearly uniform h such that the solution is in the asymptotic range of convergence in h. This may be accomplished by performing uniform h-refinements of some very coarse initial mesh until the asymptotic range is reached. The resulting grid has N_0 elements with sizes h_i, degrees p_i and *a posteriori* error estimates η_i, and approximate solution u_0. The results in Section 5 begin with $p=1$ and assume the initial grid is sufficiently fine in h.

The second step is to perform adaptive h-refinement to reach an intermediate error tolerance $\gamma\tau$ where γ is a given parameter. In the references, γ is in the range $5-10$, usually 6 in the numerical results. This intermediate grid is created by computing a desired number of children for each element T_i by the formula

$$n_i = \left[\frac{\Lambda_i^2 N_i h_i^{2\mu_i}}{p_i^{2(m_i-1)} \eta_i^2}\right]^{\frac{1}{\beta\mu_i+1}} \tag{11}$$

where $N_I = \sum n_i$ is the number of elements in the intermediate grid, m_i is the local regularity of the solution, $\mu_i = \min(p_i, m_i - 1)$, $\eta_I = \gamma \tau \|u_0\|_E$, $\beta = 1$ for 2D problems, $\eta_0^2 = \sum \eta_i^2$ and

$$\Lambda_i = \frac{\eta_i \Lambda}{\eta_0}$$

where

$$\Lambda = \frac{\eta_0 p_i^{m_i - 1}}{h_i^{\mu_i}}$$

See any of the above references for the derivation of this formula. It is based on the *a priori* error estimate in Equation 5. Inserting the expression for Λ_i into Equation 11 and using $\beta = 1$ we arrive at

$$n_i = \left(\frac{\eta_i^2 N_I}{\eta_I^2}\right)^{\frac{1}{\mu_i + 1}}$$

N_I is not known at this point, since it is the sum of the n_i. Successive iterations are used to solve for n_i and N_I simultaneously. We use 5 iterations, which preliminary experiments showed to be sufficient (convergence was usually achieved in 3 or 4 iterations). Once the n_i have been determined, we perform $\lfloor 0.5 + \log_2 n_i \rfloor$ uniform h-refinements (bisections) of each element T_i to generate approximately n_i children, and solve the discrete problem on the intermediate grid.

The third step is to perform adaptive p-refinement to reduce the error to the desired tolerance τ. The new degree for each element is given by

$$\hat{p}_i = p_i \left(\frac{\eta_{I,i} \sqrt{N_I}}{\eta_T}\right)^{\frac{1}{m_i - 1}}$$

where $\eta_{I,i}$ is the *a posteriori* error estimate for element T_i of the intermediate grid and $\eta_T = \tau \|u_0\|_E$. Again, the formula is a simple reduction of the equations derived in the references. p-refinement is performed to increase the degree of each element T_i to $\hat{p}_i$, and the discrete problem is solved on the final grid.

In the results of Section 5, if $n_i < 2$ or $\hat{p}_i < p_i$ then refinement is not performed. Also, to avoid excessive refinement, the number of h-refinements done to any element in step 2 and number of p-refinements in step 3 is limited to 3.

The strategy of performing all the h-refinement in one step and all the p-refinement in one step is adequate for low accuracy solutions (e.g. 1%), but is not likely to work well with high accuracy solution (e.g. 10^{-8} relative error)[25]. We extend the Texas 3 Step strategy to high accuracy by cycling through steps 2 and 3 until the final tolerance τ_{final} is met. τ in the algorithm above is now the factor by which one cycle of steps 2 and 3 should reduce the error. Toward this end, before step 2 the error estimate η_0 is computed for the current grid. The final (for this cycle) and intermediate targets are now given by $\eta_T = \tau \eta_0$ and $\eta_I = \gamma \eta_T$. In the results of Section 5 we use $\tau = 0.1$ and $\gamma = 6$. For the local regularity m_i we use the same routine as the APRIORI strategy (Section 4.1).

4.6 Alternate h and p

This strategy, which will be referred to as ALTERNATE, is a variation on T3S that is more like the algorithm of Figure 1. The difference from T3S is that instead of predicting the number of refinements needed to reduce the error to the next target, the usual adaptive refinement is performed until the target is reached. Thus in step 2 all elements with an error indicator larger than $\eta_i / \sqrt{N_0}$ are h-refined. The discrete problem is solved and the new error estimate compared to η_i. This is repeated until the error estimate is smaller than η_i. Step 3 is similar except adaptive p-refinement is performed and the target is η_T. Steps 2 and 3 are repeated until the final error tolerance is achieved.

4.7 Nonlinear Programming

Patra and Gupta [26] proposed a strategy for hp mesh design using nonlinear programming. We refer to this strategy as NLP. They presented it in the context of quadrilateral elements with one level of hanging nodes, i.e., an element edge is allowed to have at most one hanging node. Here it is modified slightly for newest node bisection of triangles with no hanging nodes. This is another approach that does not strictly follow the algorithm in Figure 1.

Given a current grid with triangles $\{T_i\}$, degrees $\{p_i\}$, h-refinement levels $\{l_i\}$, error estimates $\{\eta_i\}$, and element diameters

$$h_i = \sqrt{\frac{1}{2}}^{l_i} H_{0,i}$$

where $H_{0,i}$ is the diameter of the element in the initial grid that contains T_i, the object is to determine new mesh parameters $\{\hat{p}_i\}$ and $\{\hat{l}_i\}$, $i = 1..N_T$, by solving an optimization problem. The new grid is obtained by refining T_i $\hat{l}_i - l_i$ times (or coarsening if $\hat{l}_i < l_i$) and assigning degree $\hat{p}_i$ to the $2^{\hat{l}_i - l_i}$ children. The size of the children of T_i is

$$\hat{h}_i = \sqrt{\frac{1}{2}}^{\hat{l}_i} H_{0,i}.$$

There are two forms of the optimization problem, which can be informally stated as 1) minimize the number of degrees of freedom (or some other measure of grid size) subject to the error being less than a given tolerance and other constraints, and 2) minimize the error subject to the number of degrees of freedom being less than a given limit and other constraints. We will only consider the first form here; the second form simply reverses the objective function and constraint.

Computationally, the square of the error is approximated by $\sum_{i=0}^{N_T} \hat{\eta}_i^2$ where $\hat{\eta}_i$, to be defined later, is an estimate of the error in the refined grid over the region covered by T_i. The number of degrees of freedom associated with a triangle of degree p is taken to be 3/6 (one for each vertex with an average of six triangles sharing a vertex) plus 3(p − 1)/2 (p − 1 for each edge with two triangles sharing an edge) plus (p − 1)(p − 2)/2 (for the interior), which is $p^2/2$. Thus the number of degrees of freedom over the children of T_i is $2^{\hat{l}_i - l_i} \hat{p}_i^2 / 2$.

11

We can now formally state the optimization problem as

$$\underset{\{\hat{l}_i\},\{\hat{p}_i\}}{\text{minimize}} \sum_{i=1}^{N_T} 2^{\hat{l}_i - l_i} \frac{\hat{p}_i^2}{2} \tag{12}$$

$$\text{s.t.} \sum_{i=1}^{N_T} \hat{\eta}_i^2 < \hat{\tau}^2 \tag{13}$$

$$\hat{l}_j - 1 \leq \hat{l}_i \leq \hat{l}_j + 1 \quad \forall j \text{ such that } T_j \text{ shares an edge with } T_i \tag{14}$$

$$0 \leq \hat{l}_i \leq l_{max} \tag{15}$$

$$1 \leq \hat{p}_i \leq p_{max} \tag{16}$$

$$l_i - \Delta l_{dec} \leq \hat{l}_i \leq l_i + \Delta l_{inc} \tag{17}$$

$$p_i - \Delta p_{dec} \leq \hat{p}_i \leq p_i + \Delta p_{inc} \tag{18}$$

where $\hat{\tau}$ is the error tolerance for this refinement phase. We use $\hat{\tau} = \eta/4$ where η is the global error estimate on the current grid. The divisor 4 is arbitrary and could be replaced by some other value. In practice, Equation 13 is divided through by $\hat{\tau}^2$ so that the numbers are O(1). Equation 14 is a necessary condition for compatibility of the grid (in [26] it enforces one level of hanging nodes). It is not a sufficient condition, however any violations of compatibility while this condition is met are cases where only one triangle of a compatibly divisible pair was refined, and it is a slight adjustment to the optimal solution to also refine the other one to maintain compatibility. Equation 15 insures that coarsening does not go beyond the initial grid, and that the refinement level of an element does not exceed a prescribed limit l_{max}. Similarly, Equation 16 insures that element degrees do not go below one or exceed a prescribed limit p_{max}. Also, because many quantities are only approximate, it is wise to limit the amount of change that occurs to any element during one phase of refinement. Equations 17 and 18 restrict the amount of change that can occur at one time, i.e., restrict the amount of decrease in l and p to prescribed limits Δl_{dec} and Δp_{dec}, and the amount of increase to Δl_{inc} and Δp_{inc}. In the results in Section 5 we used $\Delta l_{dec} = 1$, $\Delta p_{dec} = 1$, $\Delta l_{inc} = 3$, and $\Delta p_{inc} = 1$.

Since the $\hat{l}_i$ and $\hat{p}_i$ are naturally integers, the optimization problem is a mixed integer nonlinear program, which is known to be NP-hard. To make the problem tractable, the integer requirement is dropped to give a nonlinear program which can be solved by one of several software packages. For the results in Section 5, we used the program ALGENCAN[2] Version 2.2.1 [5,9]. Following solution of the nonlinear program, the $\hat{l}_i$ and $\hat{p}_i$ are rounded to the nearest integer.

It remains to define $\hat{\eta}_i$, the estimate of the error in the refined grid over the region covered by T_i. Assuming approximate equality in the *a priori* error estimate of Equation 5, we have

$$\eta_i \approx C \frac{h_i^{\mu_i}}{p_i^{m_i - 1}} \|u\|_{H^m(T_i)}$$

and

$$\hat{\eta}_i \approx C \frac{\hat{h}_i^{\hat{\mu}_i}}{\hat{p}_i^{\hat{m}_i - 1}} \|u\|_{H^m(T_i)}$$

---

[2] The mention of specific products, trademarks, or brand names is for purposes of identification only. Such mention is not to be interpreted in any way as an endorsement or certification of such products or brands by the National Institute of Standards and Technology. All trademarks mentioned herein belong to their respective owners.

where m_i is the local regularity over T_i and $\mu_i = \min(p_i, m_i - 1)$. Combining these leads to

$$\hat{\eta}_i \approx \frac{\hat{h}_i^{\mu_i}}{\hat{p}_i^{m_i-1}} \frac{p_i^{m_i-1}}{h_i^{\mu_i}} \eta_i = \sqrt{\frac{1}{2}}^{\mu_i(\hat{l}_i - l_i)} \left(\frac{p_i}{\hat{p}_i}\right)^{m_i - 1} \eta_i$$

and thus the constraint in Equation 13 is

$$\sum_{i=1}^{N_T} \frac{1}{2}^{\min(p_i, m_i - 1)(\hat{l}_i - l_i)} \left(\frac{p_i}{\hat{p}_i}\right)^{2(m_i - 1)} \eta_i^2 < \hat{\tau}^2$$

in which the only remaining quantity to be determined is m_i. Patra and Gupta suggest estimating m_i by using the observed convergence rate from two grids, with a formula very similar to that used in the PRIOR2P strategy of Section 4.2, so we use the same estimate as PRIOR2P.

4.8 Another Optimization Strategy

Another strategy based on the formulation and solution of an optimization problem is given in Novotny et al. [22]. However, it turns out that 1) the optimization does not work near singularities, so *a priori* knowledge of singularities must be used to force h-refinement near singularities, and 2) for the finite element method and class of problems considered in this paper, the strategy always chooses p-refinement except for extremely large elements. Thus, this strategy is (nearly) identical to the APRIORI strategy, and will not be considered further in this paper.

4.9 Predict Error Estimate on Assumption of Smoothness

Melenk and Wohlmuth [17] proposed a strategy based on a prediction of what the error should be if the solution is smooth. We call this strategy SMOOTH-PRED.

When refining element T_i, assume the solution is locally smooth and that the optimal convergence rate is obtained. If h-refinement is performed and the degree of T_i is p_i, then the expected error on the two children of T_i is reduced by a factor of 2^{p_i} as indicated by the *a priori* error estimate of Equation 5. Thus if η_i is the error estimate for T_i, predict the error estimate of the children to obey $\gamma_h \eta_i / 2^{p_i}$ where γ_h is a user specified parameter. If p-refinement is performed on T_i, exponential convergence is expected, so the error should be reduced by some constant factor $\gamma_p \in (0,1)$, i.e., the predicted error estimate is $\gamma_p \eta_i$. When the actual error estimate of a child becomes available, it is compared to the predicted error estimate. If the error estimate is less than or equal to the predicted error estimate, then p-refinement is indicated for the child. Otherwise, h-refinement is indicated since presumably the assumption of smoothness was wrong. For the results in Section 5 we use $\gamma_h = 2$ and $\gamma_p = 0.4$.

4.10 Larger of h-Based and p-Based Error Indicators

In 1D, Schmidt and Siebert [28] proposed a strategy that uses two *a posteriori* error estimates to predict whether h-refinement or p-refinement will reduce the error more. We extend this strategy to bisected triangles and refer to it as H&P-ERREST.

The local Neumann residual error estimate given by Equations 7-10 is actually an estimate of how much the norm of the solution will change if T_i is p-refined. This is because the solution of the local problem is estimated using the p-hierarchical bases that would be added if T_i is p-refined, so it is an estimate of the actual change that would occur. Using the fact that the current space is a subspace of the refined space and Galerkin orthogonality, it can be shown that

$$\|u - \hat{u}_{hp}\|^2 = \|u - u_{hp}\|^2 - \|\hat{u}_{hp} - u_{hp}\|^2$$

where $\hat{u}_{hp}$ is the solution in the refined space. Thus the change in the solution indicates how much the error will be reduced.

A second error estimate for T_i can be computed by solving a local Dirichlet problem

$$L e_i = f - L u_{hp} \quad \text{in } T_i \cup T_i^{mate} \tag{19}$$
$$e_i = g_D - u_{hp} \quad \text{on } \partial(T_i \cup T_i^{mate}) \cap \partial\Omega_D \tag{20}$$
$$B e_i = g_N - B u_{hp} \quad \text{on } \partial(T_i \cup T_i^{mate}) \cap \partial\Omega_N \tag{21}$$
$$e_i = 0 \quad \text{on } \partial(T_i \cup T_i^{mate}) \setminus \partial\Omega_D \setminus \partial\Omega_N \tag{22}$$

where T_i^{mate} is the element that is refined along with T_i in the newest node bisection method [20]. The solution to this problem is approximated by an h-refinement of the two elements using only the new basis functions. The error estimate obtained by taking the norm of this approximate solution is actually an estimate of how much the solution will change, or the error will be reduced, if h-refinement is performed.

Schmidt and Siebert divide the two error estimates by the associated increase in the number of degrees of freedom to obtain an approximate error reduction per degree of freedom. In addition or instead, one of the error estimates can be multiplied by a user specified constant to bias the refinement toward h- or p-refinement. In the results of Section 5 the p-based error estimate is multiplied by 2.

The type of refinement that is used is the one that corresponds to the larger of the two modified error estimates.

4.11 Legendre coefficient strategies

There are three hp-adaptive strategies that are based on the coefficients in an expansion of the solution in Legendre polynomials. In one dimension, the approximate solution in element T_i with degree p_i can be written

$$u_i(x) = \sum_{j=0}^{p_i} a_j P_j(x)$$

where P_j is the jth degree Legendre polynomial scaled to the interval of element T_i.

Mavriplis[16] estimates the decay rate of the coefficients by a least squares fit of the the last four coefficients a_j to $Ce^{-\sigma j}$. Elements are refined by p-refinement where $\sigma > 1$ and by h-refinement where $\sigma \leq 1$. We refer to this strategy as COEF-DECAY. When four coefficients are not available, we fit to whatever is available. If only one coefficient is available, we use p-refinement.

Houston et al. [15] present the other two approaches which use the Legendre coefficients to estimate the regularity of the solution. One approach estimates the regularity using the root test yielding

$$m_i = \frac{\log \frac{2p_i+1}{2a_{p_i}^2}}{2\log p_i}.$$

If $p_i = 1$, use p-refinement. Otherwise, use p-refinement if $p_i \leq m_i - 1$ and h-refinement if $p_i > m_i - 1$. We refer to this strategy as COEF-ROOT.

They also present a second way of estimating the regularity from the Legendre coefficients using the ratio test. However, they determined the ratio test is inferior to the root test, so it will not be considered further in this paper.

Both Mavriplis and Houston et al. presented the strategies in the context of one dimension and use the Legendre polynomials as the local basis so the coefficients are readily available. In [15] it is extended to 2D for rectangular elements with a tensor product of Legendre polynomials, and the regularity is estimated in each dimension separately, so the coefficients are still readily available. Eibner and Melenk [12] extended the COEF-DECAY strategy to quadrisected triangles with an orthogonal polynomial basis. In this study we are using triangular elements which have a basis that is based on Legendre polynomials [33]. In this basis there are $3 + \max(j - 2, 0)$ basis functions of exact degree j over an element, so we don't have a single Legendre polynomial coefficient to use. Instead, for the coefficients a_j we use the ℓ_1 norm of the coefficients of the degree j basis functions, i.e.

$$a_j = \sum_{\substack{k \text{ s.t. } \deg(\varphi_k) = j \\ \text{supp}(\varphi_k) \cap T_i = \emptyset}} |\alpha_k|$$

4.12 Reference Solution Strategies

Demkowicz and his collaborators developed an hp-adaptive strategy over a number of years, presented in several papers and books, e.g. [10, 11, 27, 31]. In its full glory, the strategy is quite complicated. Here we present only the basic ideas of the algorithm and how we have adapted it for bisected triangles (it is usually presented in the context of rectangular elements with some reference to quadrisection triangles), and refer to the references for further details. We refer to this strategy as REFSOLN-EDGE because it relies on computing a reference solution and bases the refinement decisions on edge refinements. Note that for this strategy the basic form of the hp-adaptive algorithm is different than that in Figure 1.

The algorithm is first presented for 1D elliptic problems. Given the current existing (coarse) mesh, $G_{h,p} := G_{hp}$, and current solution, $u_{h,p} := u_{hp}$, a uniform refinement in both h and p is performed to obtain a fine mesh $G_{h/2,p+1}$. The equation is solved on the fine mesh to obtain a reference solution $u_{h/2,p+1}$. The norm of the difference between the current solution and reference solution is used as the global error estimate, i.e.,

$$\eta = \|u_{h/2,p+1} - u_{h,p}\|_{H^1}$$

The next step is to determine the optimal refinement of each element. This is done by considering a p-refinement and all possible (bisection) h-refinements (i.e., all possible assignments of p to the two children of an h-refinement) that give the same increase in the number of degrees of freedom as the p-refinement. In 1D, this means that the sum of the degrees of the two children must be p+1, resulting in a total of p h-refinements and one p-refinement to be examined. For each possibility, the error decrease rate is computed as

$$\frac{|u_{h/2,p+1} - \Pi_{hp,i} u_{h/2,p+1}|^2_{H^1(T_i)} - |u_{h/2,p+1} - \Pi_{new,i} u_{h/2,p+1}|^2_{H^1(T_i)}}{N_{new} - N_{hp}}$$

where $\Pi_{hp,i} u_{h/2,p+1}$ is the projection-based interpolant of the reference solution in element T_i, computed by solving a local Dirichlet problem, and $\Pi_{new,i}$ is the projection onto the resulting elements from any one of the candidate refinements. The refinement with the largest error decrease rate is selected as the optimal refinement. In the case of h-refinement, the degrees may be increased further by a process known as following the biggest subelement error refinement path, which is also used to determine the guaranteed element rate; see [10] for details.

Elements that have a guaranteed rate larger than 1/3 the maximum guaranteed rate are selected for refinement; the factor 1/3 is arbitrary.

The 2D algorithm also begins by computing a reference solution on the globally hp-refined grid $\tilde{G}_{h/2,p+1}$. (For bisected triangles, we should use the subscript $h/2, p+1$ for the fine grid and solution, but for simplicity we will use the original notation.) Then for each edge in the grid, the choice between p- and h-refinement, the determination of the guaranteed edge rate, and the selection of edges to refine are done exactly as in 1D, except that a weighted H^1 seminorm is used instead of the more natural $H^{1/2}$ seminorm which is difficult to work with. In the case of bisected triangles, we only consider edges that would be refined by the bisection of an existing triangle.

The h-refinement of edges determines the h-refinement of elements. It remains to determine the degree of each element. As a starting point, element degrees are assigned to satisfy the minimum rule for edge degrees, using the edge degrees determined in the previous step. Then the biggest subelement error refinement path is followed to determine the guaranteed element rate and assignment of element degrees. We again refer to [10] for details. Finally, the minimum rule for edge degrees is enforced by increasing edge degrees as necessary.

A related, but simpler, approach was developed by Šolín et al. [30]. We refer to this strategy as REF-SOLN-ELEM since it also begins by computing a reference solution, $u_{h/2,p+1}$, on $G_{h/2,p+1}$, but bases the refinement on elements. The basic form of the hp-adaptive algorithm is different than that in Figure 1 for this strategy, also.

The local error estimate is given by the norm of the difference between the reference solution and the current solution,

$$\eta_i = \|u_{h/2,p+1} - u_{h,p}\|_{H^1(T_i)}$$

and the elements with the largest error estimates are refined. If T_i is selected for refinement, let $p_0 = \lfloor (p_i +1)/2 \rfloor$ and consider the following options:

- p-refine T_i to degree $p_i + 1$,
- p-refine T_i to degree $p_i + 2$,
- h-refine T_i and consider all combinations of degrees p_0, p_0+1 and p_0+2 in the children.

In all cases the minimum rule is used to determine edge degrees. In [30], quadrisection of triangles is used leading to 83 options to consider. With bisection of triangles, there are only 11 options. Also, since the object of dividing by two to get p_0 is to make the increase in degrees of freedom from h-refinement comparable to that of p-refinement, we use $p_0 = \lfloor (p_i+1)/2 \rfloor$ since there are only two children instead of four. Šolín et al. allow an unlimited number of hanging nodes, so they have no issue of how to assign the degrees of children that were created to maintain compatibility or one level of hanging nodes. For the newest node bisection of triangles algorithm, we assign $\lfloor (p+1)/2 \rfloor$ to both children of a triangle of degree p that is refined only for the sake of compatibility.

For each candidate, the standard H^1 projection $\Pi_{\text{candidate}}^{H^1(T_i)}$ of $u_{h/2,p+1}$ onto the corresponding space is performed, and the projection error in the H^1 norm, $\zeta_{\text{candidate}}$, is computed,

$$\zeta_{\text{candidate}} = \|u_{h/2,p+1} - \Pi_{\text{candidate}}^{H^1(T_i)} u_{h/2,p+1}\|_{H^1(T_i)}$$

as well as the projection error of the projection onto T_i, ζ_i.

The selection of which candidate to use is not simply the candidate with the smallest projection error [29]. Let N_i be the number of degrees of freedom in the space corresponding to T_i, and $N_{\text{candidate}}$ the number of degrees of freedom in the space corresponding to a candidate. For simplicity, when computing N_i and $N_{candidate}$ we apply the minimum rule for edge degree ignoring the degrees of the neighbors of T_i, e.g. $N_i = (p_i+1)(p_i+2)/2$ regardless of what the actual edge degrees of T_i are.

Candidates with $\zeta_{candidate} > \zeta_i$ are discarded. We also discard any of the h-refined candidates for which the degrees are both greater than p_i since the reference solution is (locally) in that space. Let n be the number of remaining candidates. Compute the average and standard deviation of the base 10 logarithms of the ζ's

$$\bar{\zeta} = \frac{1}{n} \sum_{candidates} \log \zeta_{candidate}$$

$$\sigma = \sqrt{\frac{1}{n} \sum_{candidates} (\log \zeta_{candidate})^2 - \bar{\zeta}^2}$$

Finally, to determine which candidate to use, select an above-average candidate with the steepest error decrease, i.e., from among the candidates with $\log \zeta_{candidate} < \bar{\zeta} + \sigma$ and $N_{candidate} > N_i$, select the candidate that maximizes

$$\frac{\log \zeta_i - \log \zeta_{candidate}}{N_{candidate} - N_i} \tag{23}$$

Following the refinement that is indicated by the selected candidate, the minimum rule for edge degrees is applied.

This algorithm can be modified slightly to bias the refinement towards or away from p-refinement to improve the performance. Given a parameter p_{bias}, multiply the value from Equation 23 by it for all the p-refinement candidates. $p_{bias} > 1$ will bias the refinement toward doing p-refinement, and $p_{bias} < 1$ will bias the refinement toward doing h-refinement. For the results in Section 5 we use $p_{bias} = 2$ for most problems, and $p_{bias} = 4$ for the analytic, mild wave front and both peak problems, which are the easiest problems.

5 Numerical Results

This section contains the results of a numerical experiment to compare the hp-adaptive strategies' performance on a suite of 21 test problems with various difficulties that adaptive refinement should locate. The primary criteria for comparing the strategies is the convergence of the relative error in the energy norm as a function of the number of degrees of freedom, N. The results for each problem are given in Sections 5.1-5.21, and summary results for comparison of the strategies are given in Section 5.23. We also give some indication of the relative amount of time required to obtain the solution in Section 5.22.

The full details of the test problems can be found in a separate report [21]. Here we just give a brief description of each problem and an image of the solution, both as a color map and as a surface in perspective. Recall that Poisson's equation is $u_{xx} + u_{yy} = f(x,y)$ and Laplace's equation is Poisson's equation with $f=0$.

Each problem is solved with each hp strategy using the hp-adaptive algorithm of Section 3, except for those strategies that dictate using a variation on that algorithm, as indicated in Section 4. To examine the convergence of the error as a function of N, each problem is solved using each strategy several times with different values of the termination tolerance τ. The relative energy norm of the error and N are recorded at the end of each run to give a set of points for the convergence data. In most cases we used $\tau = 0.1, 0.05, 0.02, 0.01, 0.005 \ldots 2 \times 10^{-8}, 10^{-8}$, although some of the more difficult problems required ending the sequence earlier.

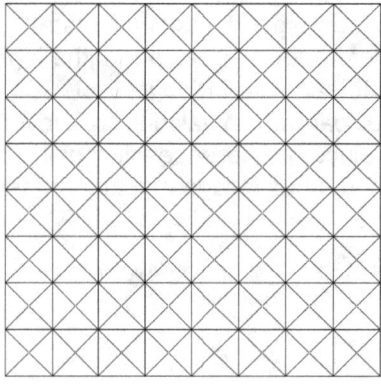

Figure 2: The initial grid for problems on square domains.

The initial grid for problems on a square domain is shown in Figure 2. The initial grid for the reentrant corner problems is obtained by removing the unneeded elements from the grid in Figure 2. The initial grid for the battery problem is shown in Section 5.11.

In the following subsections for each problem, we present the following results of the computations.

We begin with a sample grid for each strategy, to show the wide variation in the different strategies' choice between h and p refinement. In all of the grid images, the color indicates the degree of the polynomial over each element. To obtain the grid, we pick one particular value of the termination relative error tolerance τ and run each strategy to that tolerance. In cases where there is strong h-refinement in a small area we also zoom in on that area to show the detail at the fine level.

Second we present a plot of N vs. the error on a log-log scale. These graphs have a curvature indicating the exponential rate of convergence. The black circles and connecting lines show the convergence data obtained by solving the problem with a sequence of termination tolerances. Points that were obvious outliers were omitted. The red and green curves are exponential least squares fits to the data. According to Equation 6 the error should converge like $Ae^{BN^{1/3}}$. The red curve is a least squares fit to this form. As will be seen, this fit is not always close to the data. Often the data exhibit exponential convergence, but with a different exponent on N than 1/3. The green curve is a least squares fit to the form Ae^{BN^C}. This 3-parameter least squares fit will be the primary means of comparing the performance of the strategies.

Following the individual strategy convergence plots is a composite plot containing the 3-parameter least squares fit curves of all strategies on a single graph.

Some papers on hp-adaptive refinement present the convergence plots using a cube root of N vs. logarithm of error scale. This is because, if the error converges like $Ae^{BN^{1/3}}$, then the convergence plot will be a straight line using this scale. To illustrate this, we present the cube root vs. log plots for one problem, the L-shaped domain problem (Section 5.4), along with the 2-parameter least squares fit.

The parameters obtained by the 3-parameter and 2-parameter least squares fits are given in tables. In the 3-parameter fit, C (the power on N, theoretically 1/3) indicates the curvature of the curve on a log-log plot. Very small values indicate the exponential nature of the convergence is weak. Larger values indicate a larger curvature, which asymptotically gives faster convergence rates. In the 2-parameter fit, B is the slope of the line on the cube root vs. log convergence plot. Strategies with smaller values of B (larger magnitude, since B is negative) will have steeper slopes, and asymptotically be the better strategies.

Figure 3: Computation of the factor by which N for a particular strategy is larger than N for the best strategy. In this illustration, for an accuracy of 10^{-6} the factor for ALTERNATE is $53730/7787 \approx 6.90$.

The performance of the strategies are compared on each problem at two accuracy requirements. For low accuracy, which is typical in engineering applications, we use 10^{-2}, or 1% relative error, for most problems. For high accuracy, which is of interest mathematically and useful in some scientific applications, we use 10^{-6} for most problems.

To compare the strategies, consider the 3-parameter least squares fit. For each strategy, compute the value of N that gives the desired accuracy according to the formula for the 3-parameter least squares fit, as illustrated in Figure 3. Let N_{best} be the minimum such value over all the strategies. For each strategy compute the factor by which N is larger than the best strategy, $N_{strategy}/N_{best}$. For example, in Figure 3 the factor for ALTERNATE is $53730/7787 \approx 6.90$. The final tables of each subsection contain these factors at low and high accuracy, with the accuracy requirement given in the caption. The strategies are reordered by increasing value of the factor, implicitly giving the rank (first, second, etc.) of each strategy.

These computations were performed using the adaptive finite element code PHAML Version 1.8.1 [18] on a single processor. During the period of this investigation there were changes to the available hardware and software, but we do not believe any of these changes would effect the outcome of these computations, except in Section 5.22 where a consistent computational environment is used. The computers were 32-bit and 64-bit x86-class computers operating under CentOS 5.x distributions of Linux. PHAML was compiled with the Intel Fortran compiler.

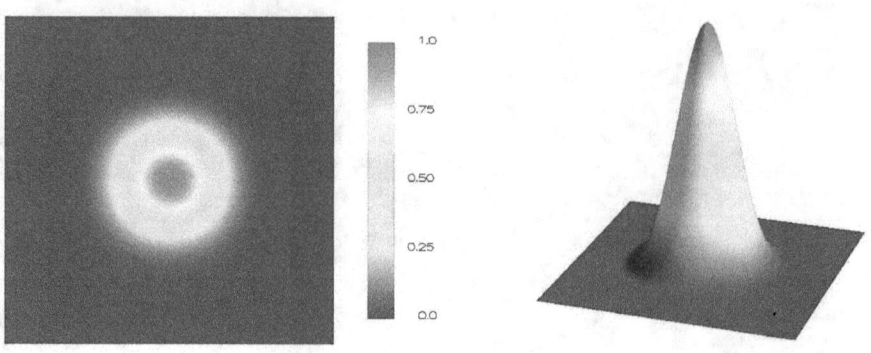

Figure 4: The solution of the analytic problem.

5.1 Analytic Solution

The analytic problem in [21] is Poisson's equation on the unit square with Dirichlet boundary conditions. The solution is the polynomial

$$2^{4p} x^p (1-x)^p y^p (1-y)^p$$

with $p=10$. 2^{4p} is a normalization factor so that the L^∞ norm is 1.0. The purpose of this test problem is to see how the methods perform on a smooth, well-behaved problem that does not really need adaptive refinement at all. For the grid images we used $\tau = 10^{-4}$. For the APRIORI strategy, we choose to always refine by p, i.e., it is just p-adaptive refinement.

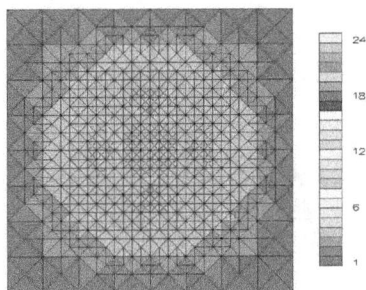

Figure 5: Example grid for the ALTERNATE strategy with the analytic problem.

Figure 8: Example grid for the COEFROOT strategy with the analytic problem.

Figure 6: Example grid for the APRIORI strategy with the analytic problem.

Figure 9: Example grid for the H&PERREST strategy with the analytic problem.

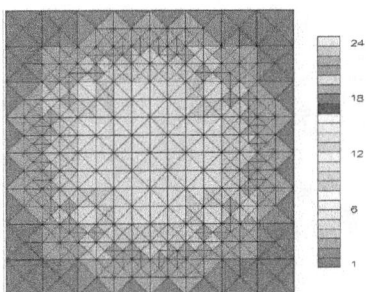

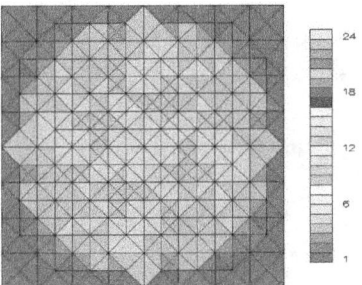

Figure 7: Example grid for the COEFDECAY strategy with the analytic problem.

Figure 10: Example grid for the NEXT3P strategy with the analytic problem.

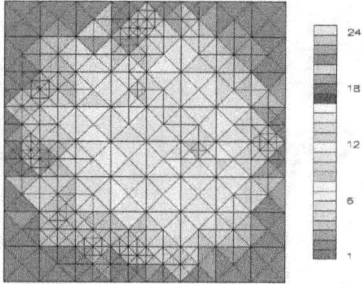

Figure 11: Example grid for the NLP strategy with the analytic problem.

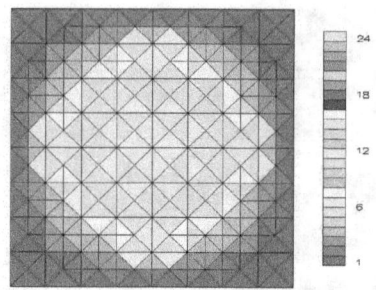

Figure 14: Example grid for the REFSOLNELEM strategy with the analytic problem.

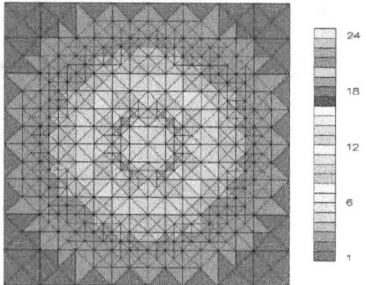

Figure 12: Example grid for the PRIOR2P strategy with the analytic problem.

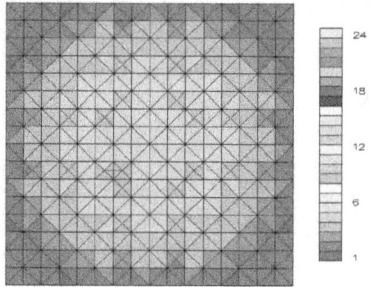

Figure 15: Example grid for the SMOOTHPRED strategy with the analytic problem.

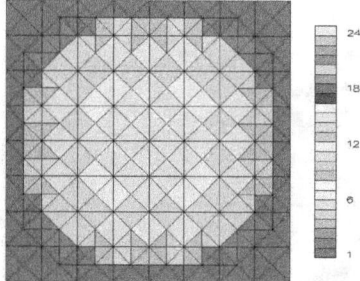

Figure 13: Example grid for the REFSOLNEDGE strategy with the analytic problem.

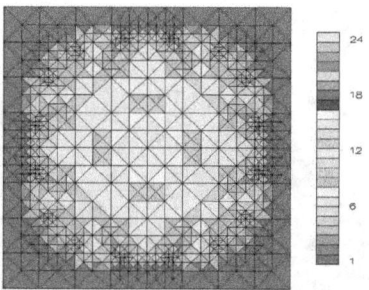

Figure 16: Example grid for the T3S strategy with the analytic problem.

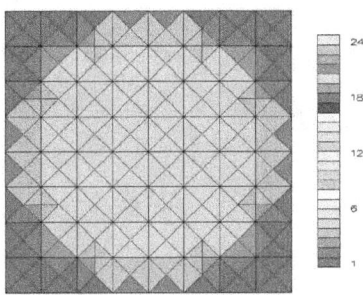

Figure 17: Example grid for the TYPEPARAM strategy with the analytic problem.

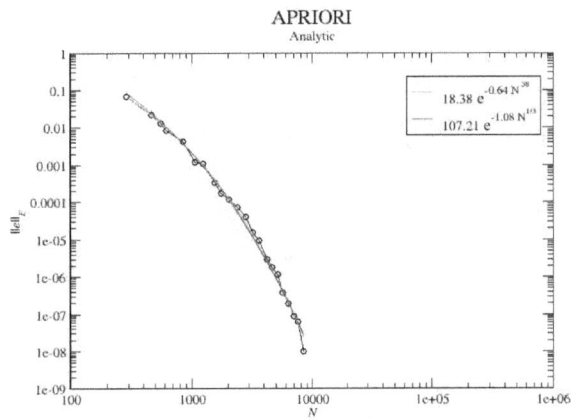

Figure 19: Log-Log plot of the convergence of the APRIORI strategy with the analytic problem.

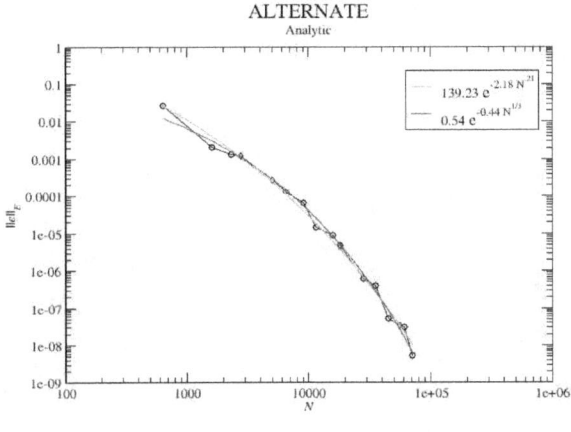

Figure 18: Log-Log plot of the convergence of the ALTERNATE strategy with the analytic problem.

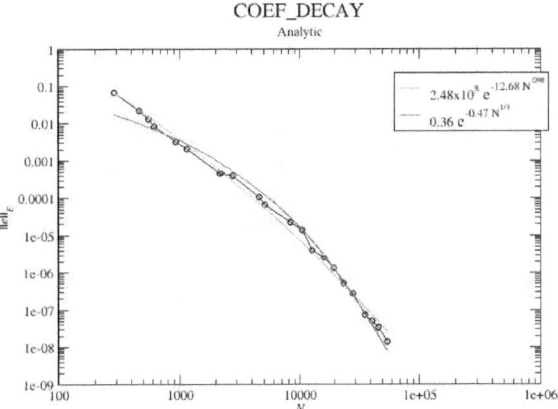

Figure 20: Log-Log plot of the convergence of the COEF-DECAY strategy with the analytic problem.

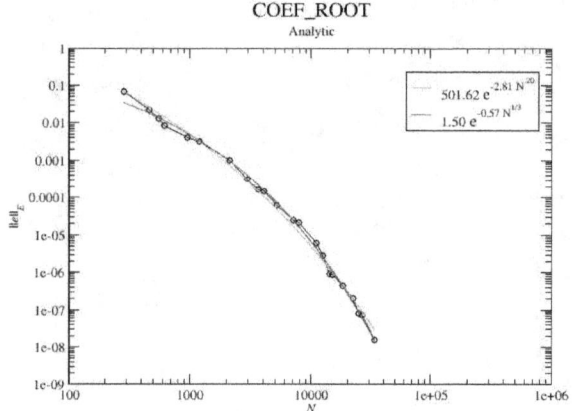

Figure 21: Log-Log plot of the convergence of the COEF-ROOT strategy with the analytic problem.

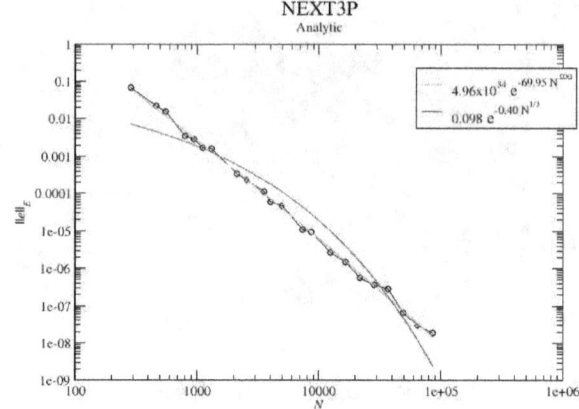

Figure 23: Log-Log plot of the convergence of the NEXT3P strategy with the analytic problem.

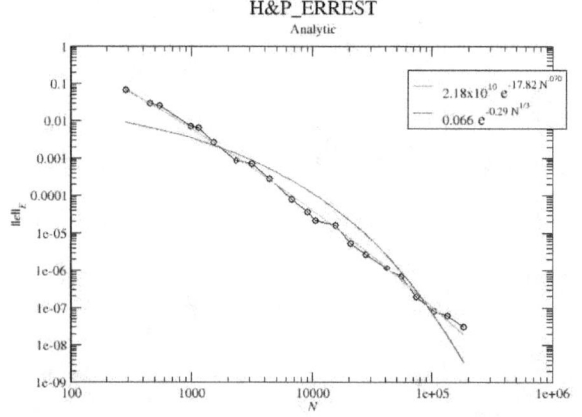

Figure 22: Log-Log plot of the convergence of the H&P-ERREST strategy with the analytic problem.

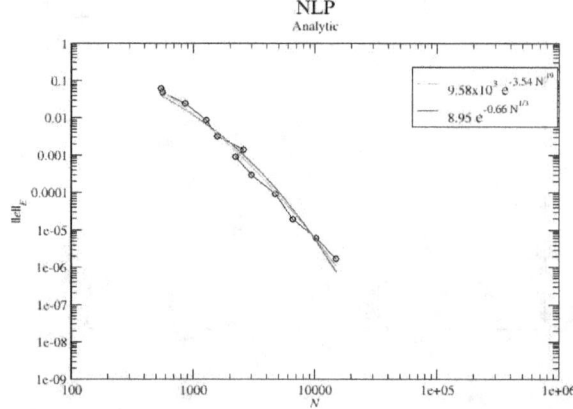

Figure 24: Log-Log plot of the convergence of the NLP strategy with the analytic problem.

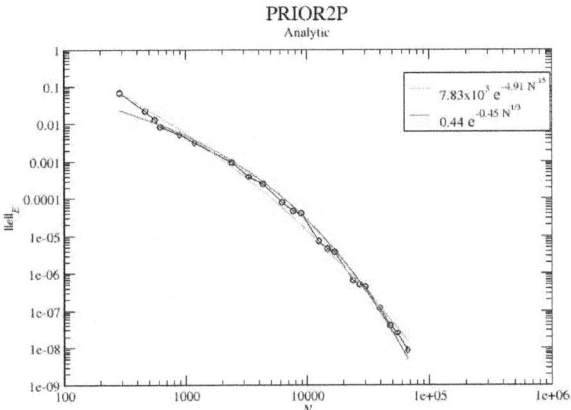

Figure 25: Log-Log plot of the convergence of the PRIOR2P strategy with the analytic problem.

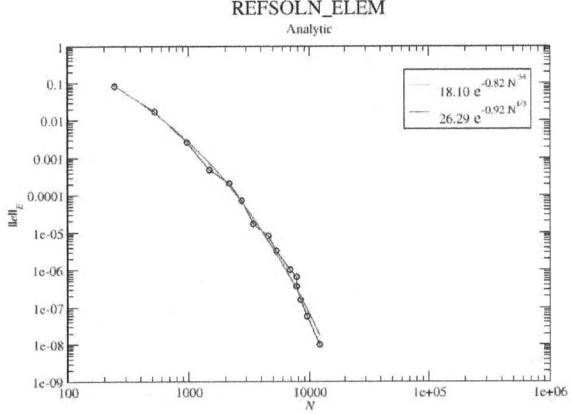

Figure 27: Log-Log plot of the convergence of the REFSOLN-ELEM strategy with the analytic problem.

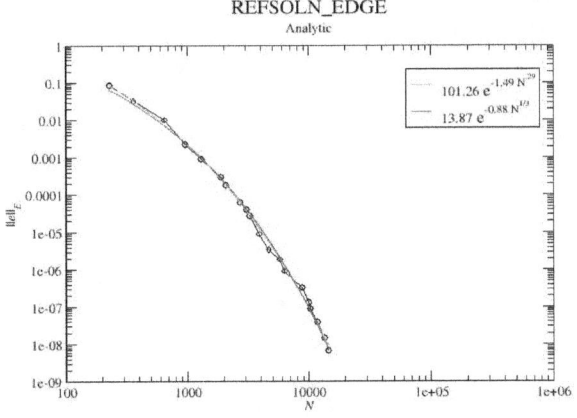

Figure 26: Log-Log plot of the convergence of the REFSOLN-EDGE strategy with the analytic problem.

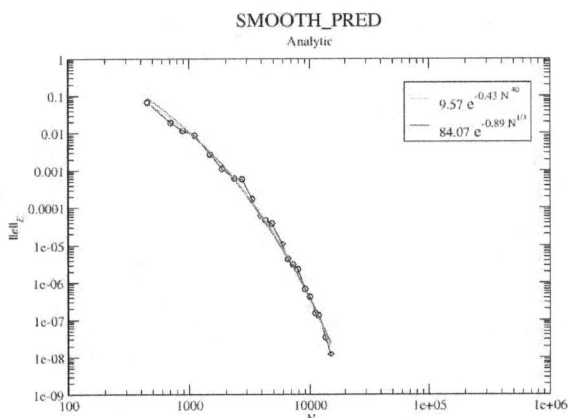

Figure 28: Log-Log plot of the convergence of the SMOOTH-PRED strategy with the analytic problem.

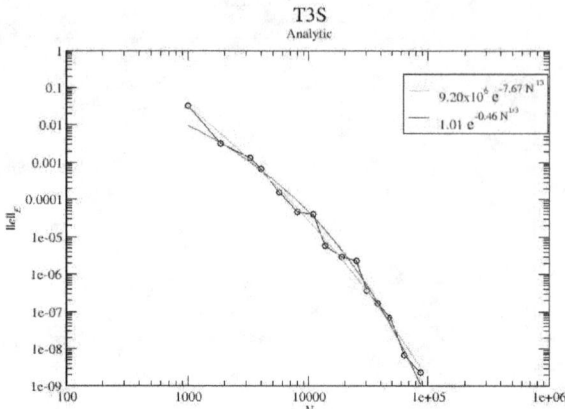

Figure 29: Log-Log plot of the convergence of the T3S strategy with the analytic problem.

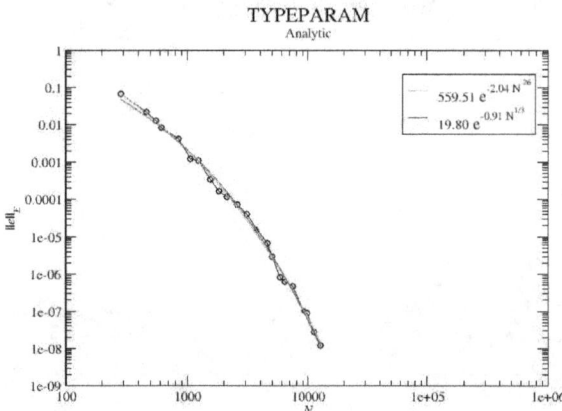

Figure 30: Log-Log plot of the convergence of the TYPEPARAM strategy with the analytic problem.

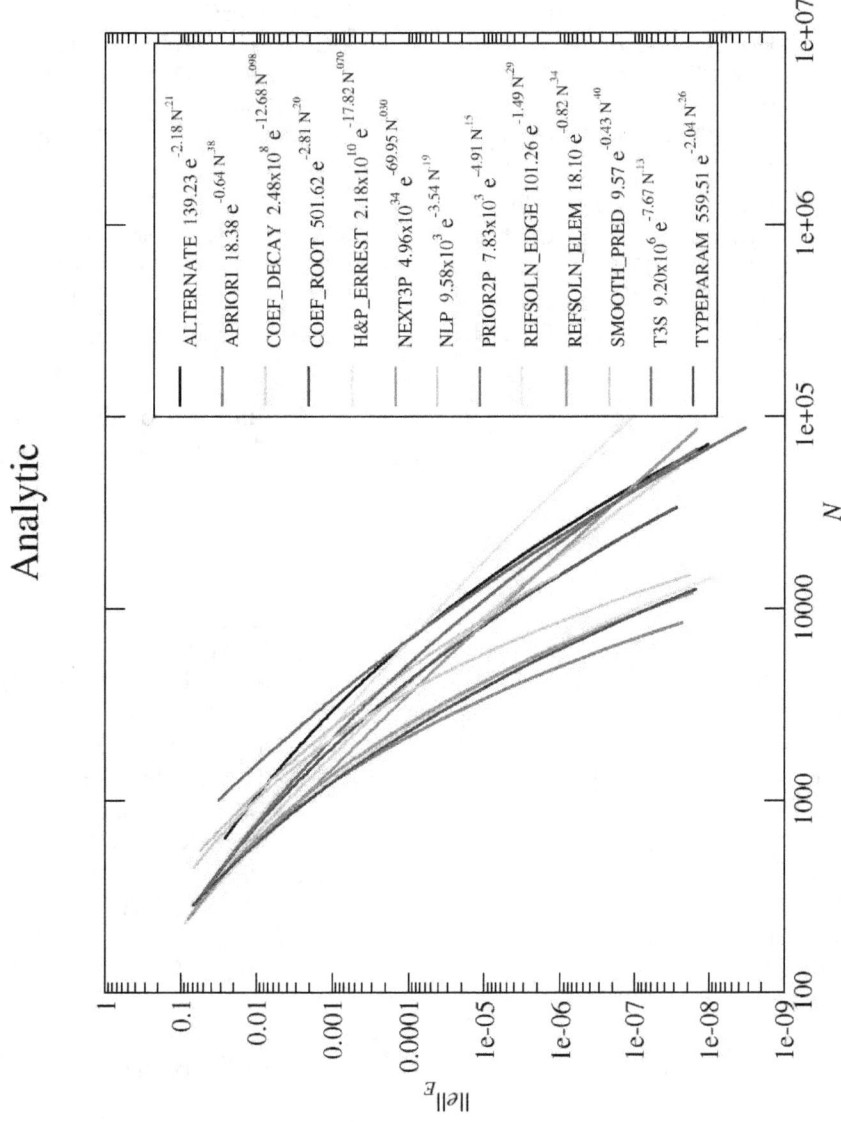

Figure 31: Log-Log plot of the convergence of all strategies with the analytic problem.

strategy	A	B	C
ALTERNATE	139.23	-2.18	0.21
APRIORI	18.38	-0.64	0.38
COEFDECAY	2.48×10^8	-12.68	0.098
COEFROOT	501.62	-2.81	0.20
H&PERREST	2.18×10^{10}	-17.82	0.070
NEXT3P	4.96×10^{34}	-69.95	0.030
NLP	9.58×10^3	-3.54	0.19
PRIOR2P	7.83×10^3	-4.91	0.15
REFSOLNEDGE	101.26	-1.49	0.29
REFSOLNELEM	18.10	-0.82	0.34
SMOOTHPRED	9.57	-0.43	0.40
T3S	9.20×10^6	-7.67	0.13
TYPEPARAM	559.51	-2.04	0.26

Table 1: Parameters of the least squares fit for $\|e_{hp}\|_E = A e^{BN_{dof}^C}$ for the analytic problem.

strategy	A	B
ALTERNATE	0.54	-0.44
APRIORI	107.21	-1.08
COEFDECAY	0.36	-0.47
COEFROOT	1.50	-0.57
H&PERREST	0.066	-0.29
NEXT3P	0.098	-0.40
NLP	8.95	-0.66
PRIOR2P	0.44	-0.45
REFSOLNEDGE	13.87	-0.88
REFSOLNELEM	26.29	-0.92
SMOOTHPRED	84.07	-0.89
T3S	1.01	-0.46
TYPEPARAM	19.80	-0.91

Table 3: Parameters of the least squares fit for $\|e_{hp}\|_E = A e^{B N_{dof}^{1/3}}$ for the analytic problem.

strategy	factor
REFSOLNEDGE	1.00
NEXT3P	1.01
TYPEPARAM	1.02
APRIORI	1.06
REFSOLNELEM	1.07
COEFDECAY	1.15
COEFROOT	1.26
PRIOR2P	1.30
H&PERREST	1.35
SMOOTHPRED	1.73
ALTERNATE	1.80
NLP	1.86
T3S	2.63

Table 2: Factor by which N is larger than the best strategy for the analytic problem at low accuracy, 1.0×10^{-2}.

strategy	factor
APRIORI	1.00
TYPEPARAM	1.23
REFSOLNELEM	1.29
REFSOLNEDGE	1.32
SMOOTHPRED	1.72
COEFROOT	3.01
NLP	3.06
COEFDECAY	3.79
NEXT3P	3.91
PRIOR2P	4.48
T3S	4.77
ALTERNATE	5.13
H&PERREST	8.77

Table 4: Factor by which N is larger than the best strategy for the analytic problem at high accuracy, 1.0×10^{-6}.

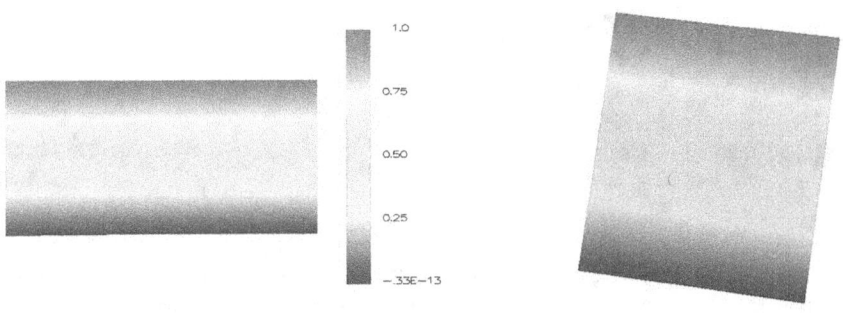

Figure 32: The solution of the nearly straight reentrant corner problem.

5.2 Reentrant Corner, Nearly Straight

For elliptic partial differential equations, a reentrant (concave) corner in the domain, with interior angle ω, causes a point singularity that behaves like r^α where r is the distance from the corner and $\alpha = \pi/\omega$. The larger ω is, the stronger the singularity. The reentrant corner problems of the next five sections are Laplace's equation with Dirichlet boundary conditions on $(-1,1) \times (-1,1)$ with a section of angle $2\pi - \omega$ removed. The solution is
$$r^\alpha \sin(\alpha\theta)$$
where $r = \sqrt{x^2 + y^2}$ and $\theta = \tan^{-1}(y/x)$.

For the nearly straight reentrant corner, $\omega = \pi + .01$. If ω was π, then there would be no reentrant corner and the solution would be linear. But with $\omega = \pi + .01$ there is a very mild singularity. For this problem, we use $\tau = 10^{-6}$ for the grid images. The APRIORI strategy refines by h if the element contains the origin and by p otherwise.

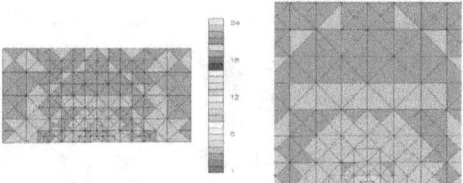

Figure 33: Example grid for the ALTERNATE strategy with the nearly straight reentrant corner problem, including details at the singularity.

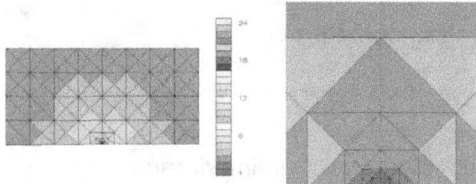

Figure 34: Example grid for the APRIORI strategy with the nearly straight reentrant corner problem, including details at the singularity.

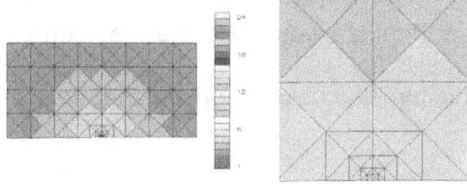

Figure 35: Example grid for the COEFDECAY strategy with the nearly straight reentrant corner problem, including details at the singularity.

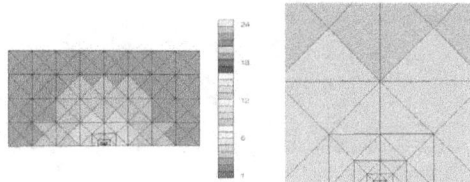

Figure 36: Example grid for the COEFROOT strategy with the nearly straight reentrant corner problem, including details at the singularity.

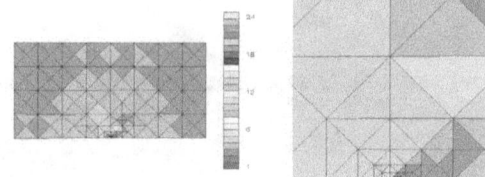

Figure 37: Example grid for the H&PERREST strategy with the nearly straight reentrant corner problem, including details at the singularity.

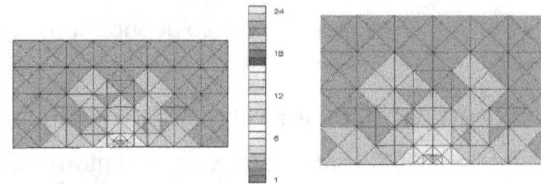

Figure 38: Example grid for the NEXT3P strategy with the nearly straight reentrant corner problem, including details at the singularity.

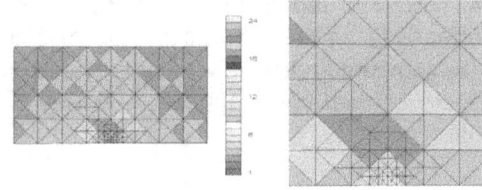

Figure 39: Example grid for the NLP strategy with the nearly straight reentrant corner problem, including details at the singularity.

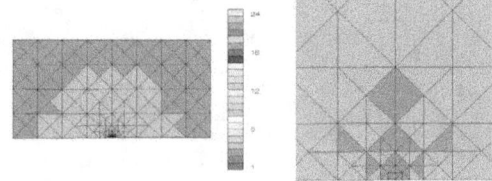

Figure 40: Example grid for the PRIOR2P strategy with the nearly straight reentrant corner problem, including details at the singularity.

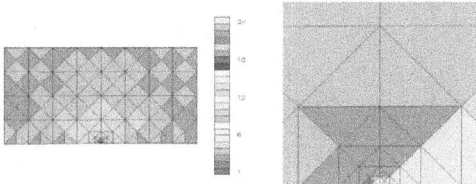

Figure 41: Example grid for the REFSOLNEDGE strategy with the nearly straight reentrant corner problem, including details at the singularity.

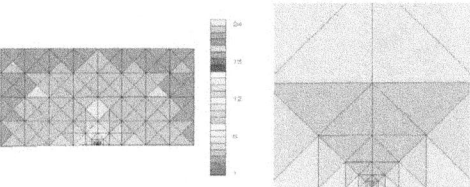

Figure 42: Example grid for the REFSOLNELEM strategy with the nearly straight reentrant corner problem, including details at the singularity.

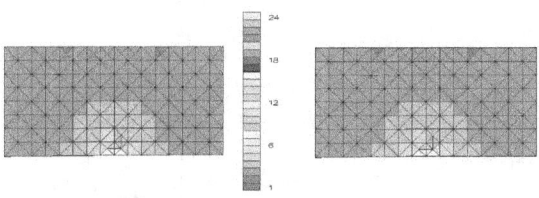

Figure 43: Example grid for the SMOOTHPRED strategy with the nearly straight reentrant corner problem, including details at the singularity.

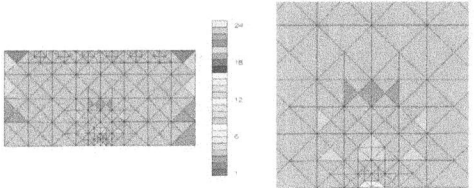

Figure 44: Example grid for the T3S strategy with the nearly straight reentrant corner problem, including details at the singularity.

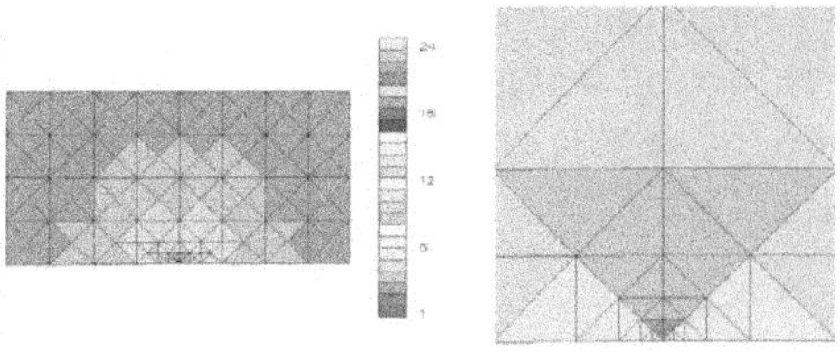

Figure 45: Example grid for the TYPEPARAM strategy with the nearly straight reentrant corner problem, including details at the singularity.

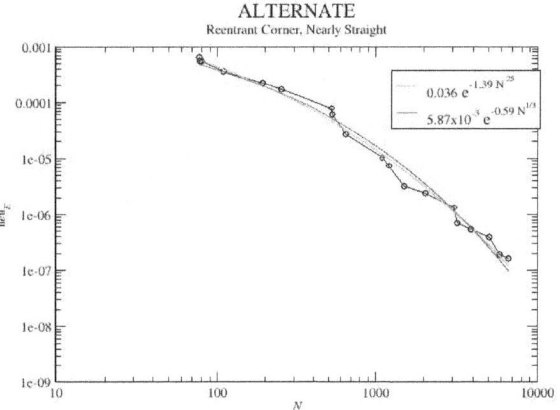

Figure 46: Log-Log plot of the convergence of the ALTERNATE strategy with the nearly straight reentrant corner problem.

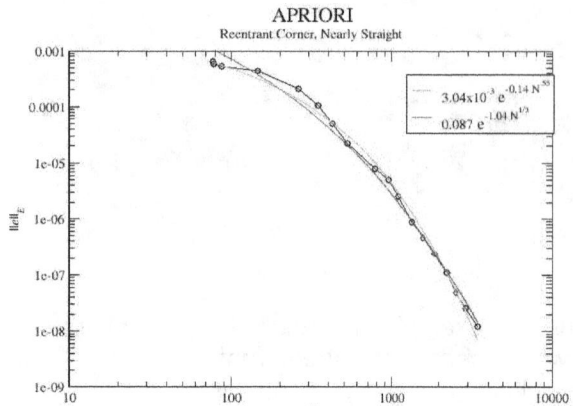

Figure 47: Log-Log plot of the convergence of the APRIORI strategy with the nearly straight reentrant corner problem.

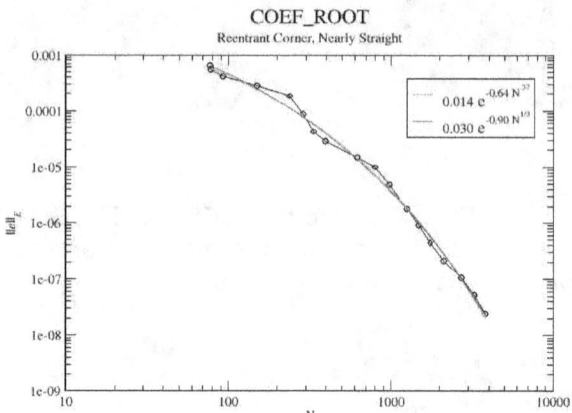

Figure 49: Log-Log plot of the convergence of the COEF-ROOT strategy with the nearly straight reentrant corner problem.

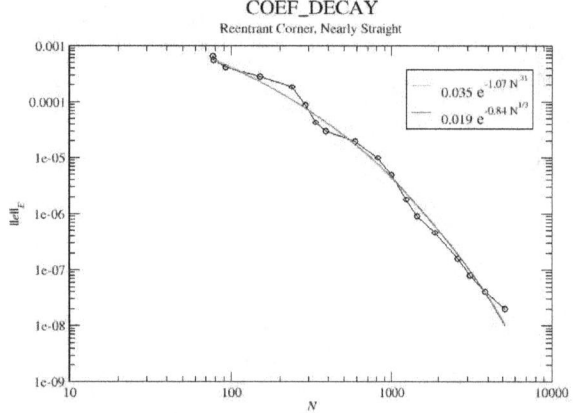

Figure 48: Log-Log plot of the convergence of the COEF-DECAY strategy with the nearly straight reentrant corner problem.

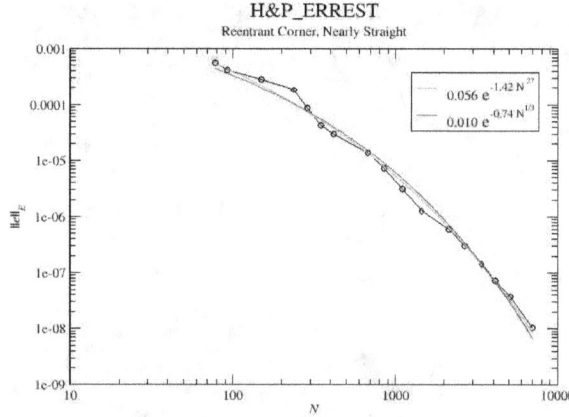

Figure 50: Log-Log plot of the convergence of the H&P-ERREST strategy with the nearly straight reentrant corner problem.

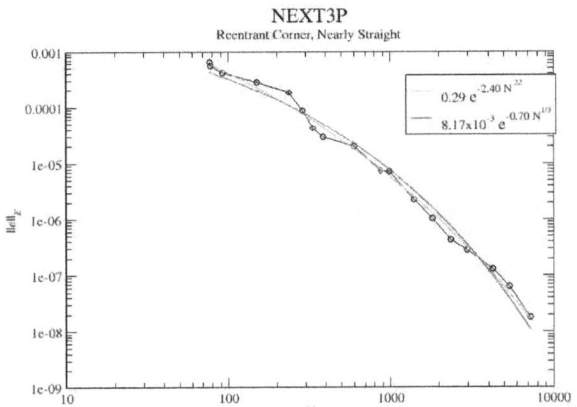

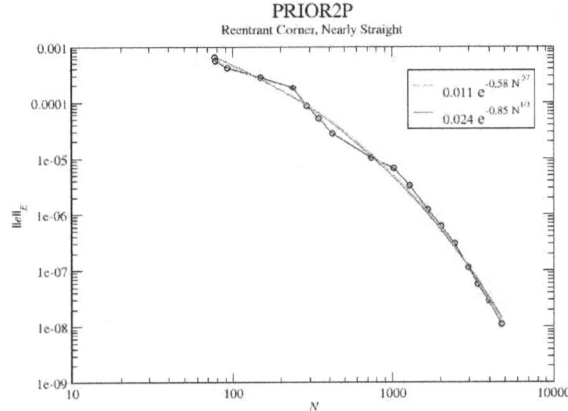

Figure 51: Log-Log plot of the convergence of the NEXT3P strategy with the nearly straight reentrant corner problem.

Figure 53: Log-Log plot of the convergence of the PRIOR2P strategy with the nearly straight reentrant corner problem.

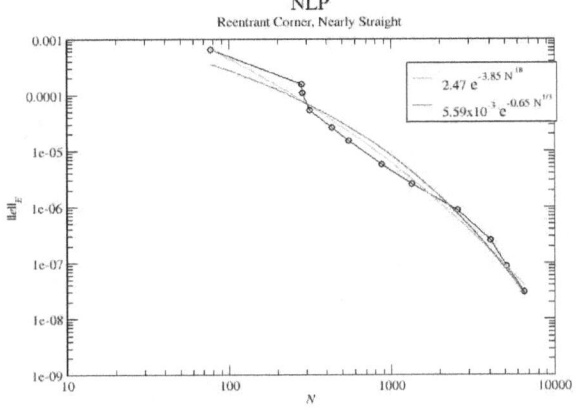

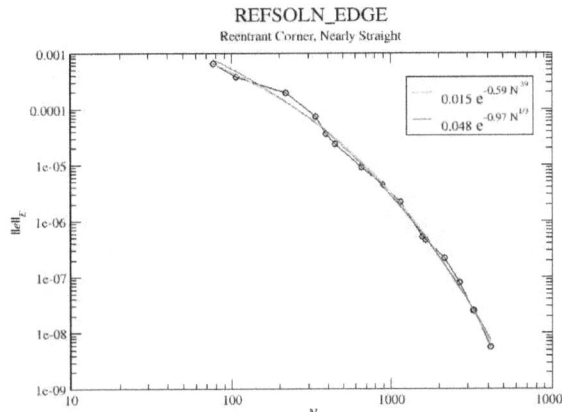

Figure 52: Log-Log plot of the convergence of the NLP strategy with the nearly straight reentrant corner problem.

Figure 54: Log-Log plot of the convergence of the REFSOLN-EDGE strategy with the nearly straight reentrant corner problem.

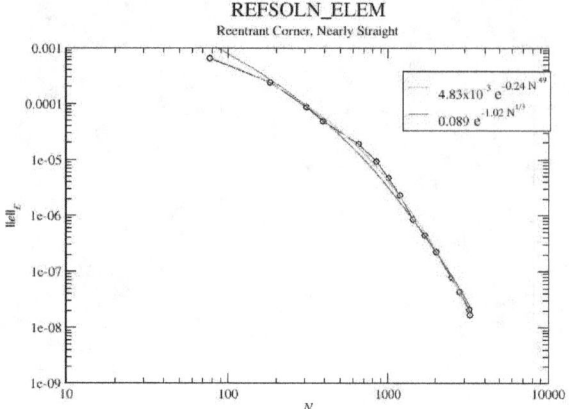

Figure 55: Log-Log plot of the convergence of the REFSOLN-ELEM strategy with the nearly straight reentrant corner problem.

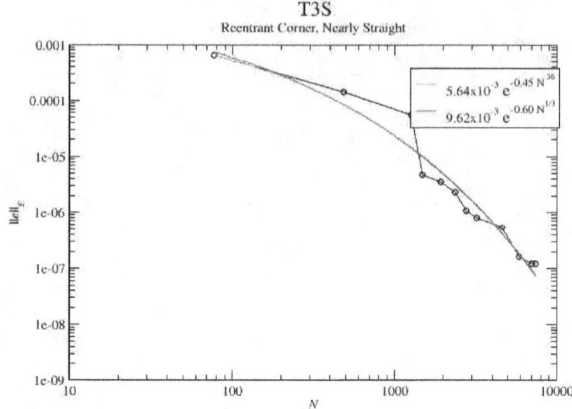

Figure 57: Log-Log plot of the convergence of the T3S strategy with the nearly straight reentrant corner problem.

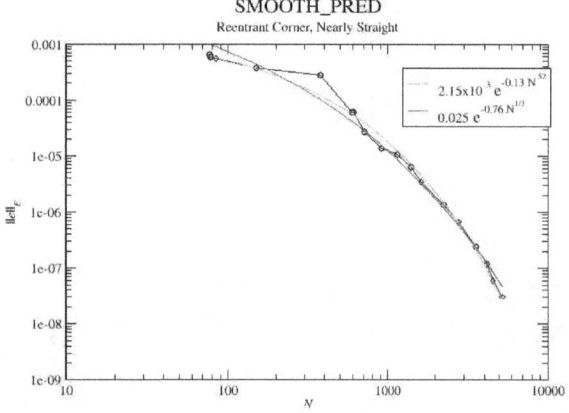

Figure 56: Log-Log plot of the convergence of the SMOOTH-PRED strategy with the nearly straight reentrant corner problem.

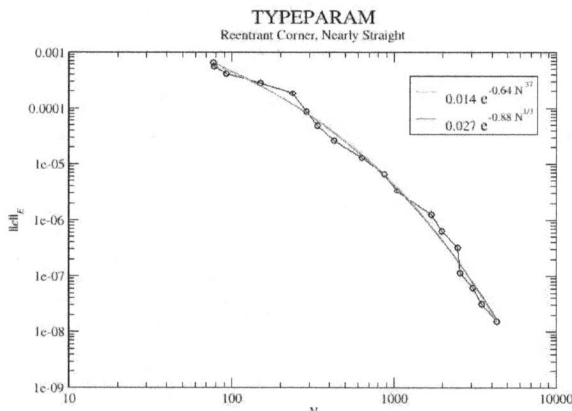

Figure 58: Log-Log plot of the convergence of the TYPEPARAM strategy with the nearly straight reentrant corner problem.

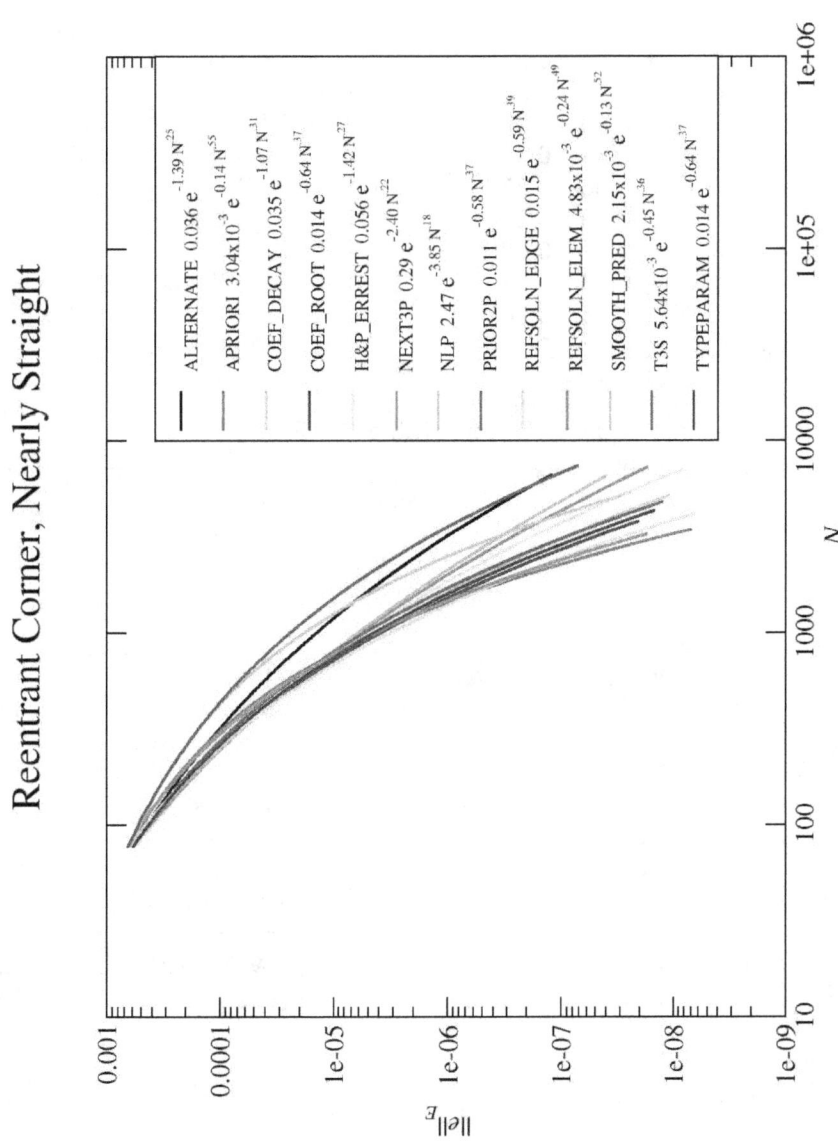

Figure 59: Log-Log plot of the convergence of all strategies with the nearly straight reentrant corner problem.

strategy	A	B	C
ALTERNATE	0.036	-1.39	0.25
APRIORI	3.04×10^{-3}	-0.14	0.55
COEFDECAY	0.035	-1.07	0.31
COEFROOT	0.014	-0.64	0.37
H&PERREST	0.056	-1.42	0.27
NEXT3P	0.29	-2.40	0.22
NLP	2.47	-3.85	0.18
PRIOR2P	0.011	-0.58	0.37
REFSOLNEDGE	0.015	-0.59	0.39
REFSOLNELEM	4.83×10^{-3}	-0.24	0.49
SMOOTHPRED	2.15×10^{-3}	-0.13	0.52
T3S	5.64×10^{-3}	-0.45	0.36
TYPEPARAM	0.014	-0.64	0.37

Table 5: Parameters of the least squares fit for $\|e_{hp}\|_E = A e^{BN^C_{dof}}$ for the nearly straight reentrant corner problem.

strategy	A	B
ALTERNATE	5.87×10^{-3}	-0.59
APRIORI	0.087	-1.04
COEFDECAY	0.019	-0.84
COEFROOT	0.030	-0.90
H&PERREST	0.010	-0.74
NEXT3P	8.17×10^{-3}	-0.70
NLP	5.59×10^{-3}	-0.65
PRIOR2P	0.024	-0.85
REFSOLNEDGE	0.048	-0.97
REFSOLNELEM	0.089	-1.02
SMOOTHPRED	0.025	-0.76
T3S	9.62×10^{-3}	-0.60
TYPEPARAM	0.027	-0.88

Table 7: Parameters of the least squares fit for $\|e_{hp}\|_E = A e^{BN^{1/3}_{dof}}$ for the nearly straight reentrant corner problem.

strategy	factor
H&PERREST	1.00
NEXT3P	1.00
COEFDECAY	1.01
NLP	1.02
COEFROOT	1.04
REFSOLNEDGE	1.04
TYPEPARAM	1.06
PRIOR2P	1.11
REFSOLNELEM	1.21
APRIORI	1.25
ALTERNATE	1.30
SMOOTHPRED	1.74
T3S	1.77

Table 6: Factor by which N is larger than the best strategy for the nearly straight reentrant corner problem at low accuracy, 1.0×10^{-4}.

strategy	factor
APRIORI	1.00
REFSOLNELEM	1.05
REFSOLNEDGE	1.07
COEFROOT	1.19
TYPEPARAM	1.27
PRIOR2P	1.36
COEFDECAY	1.36
H&PERREST	1.65
SMOOTHPRED	1.82
NEXT3P	1.91
NLP	2.14
T3S	3.00
ALTERNATE	3.10

Table 8: Factor by which N is larger than the best strategy for the nearly straight reentrant corner problem at high accuracy, 1.0×10^{-7}.

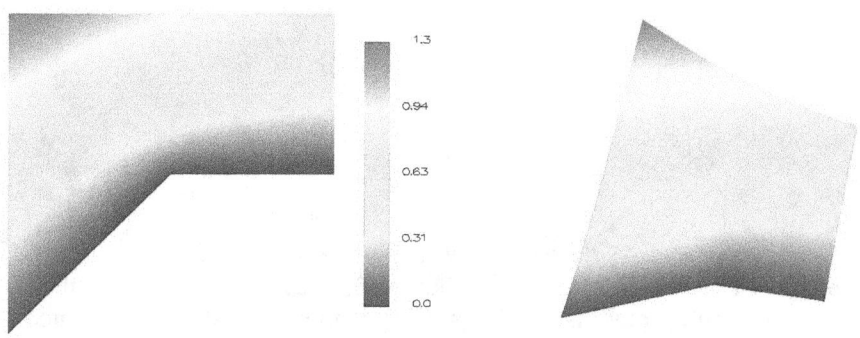

Figure 60: The solution of the wide angle reentrant corner problem.

5.3 Reentrant Corner, Wide Angle

This is the reentrant corner problem (Section 5.2) with $\omega = 5\pi/4$. $\tau = 10^{-4}$ for the grid images. The APRIORI strategy refines by h if the element contains the origin and by p otherwise.

Figure 61: Example grid for the ALTERNATE strategy with the wide angle reentrant corner problem, including details at the singularity.

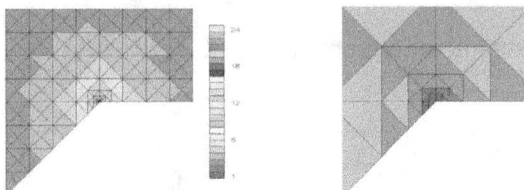

Figure 62: Example grid for the APRIORI strategy with the wide angle reentrant corner problem, including details at the singularity.

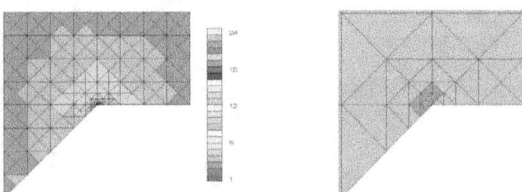

Figure 63: Example grid for the COEFDECAY strategy with the wide angle reentrant corner problem, including details at the singularity.

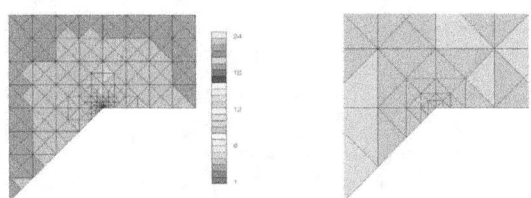

Figure 64: Example grid for the COEFROOT strategy with the wide angle reentrant corner problem, including details at the singularity.

Figure 65: Example grid for the H&PERREST strategy with the wide angle reentrant corner problem, including details at the singularity.

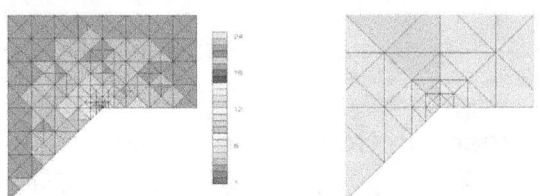

Figure 66: Example grid for the NEXT3P strategy with the wide angle reentrant corner problem, including details at the singularity.

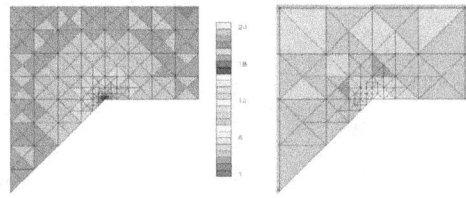

Figure 67: Example grid for the NLP strategy with the wide angle reentrant corner problem, including details at the singularity.

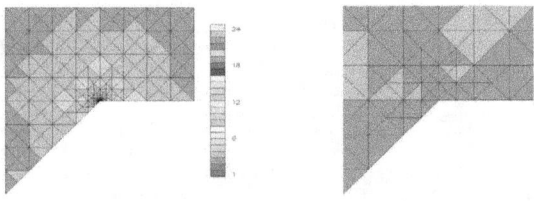

Figure 68: Example grid for the PRIOR2P strategy with the wide angle reentrant corner problem, including details at the singularity.

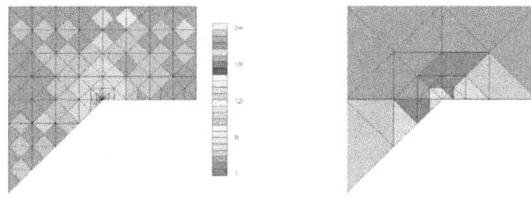

Figure 69: Example grid for the REFSOLNEDGE strategy with the wide angle reentrant corner problem, including details at the singularity.

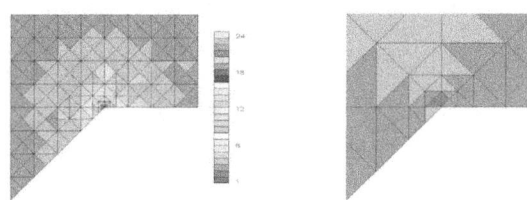

Figure 70: Example grid for the REFSOLNELEM strategy with the wide angle reentrant corner problem, including details at the singularity.

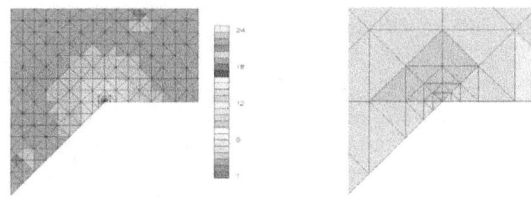

Figure 71: Example grid for the SMOOTHPRED strategy with the wide angle reentrant corner problem, including details at the singularity.

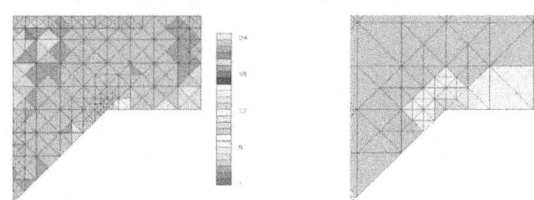

Figure 72: Example grid for the T3S strategy with the wide angle reentrant corner problem, including details at the singularity.

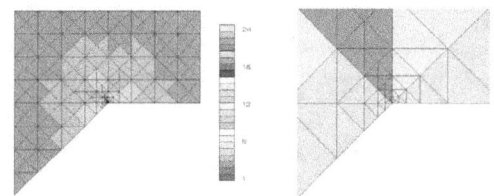

Figure 73: Example grid for the TYPEPARAM strategy with the wide angle reentrant corner problem, including details at the singularity.

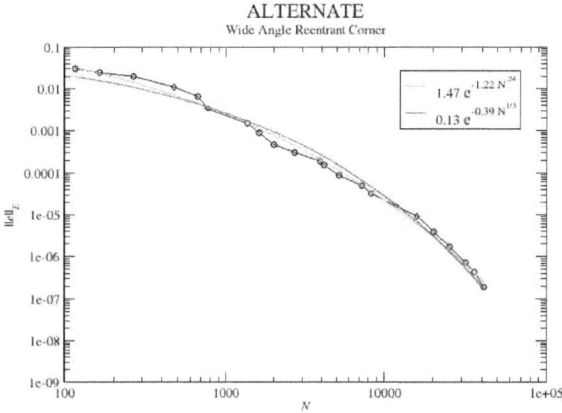

Figure 74: Log-Log plot of the convergence of the ALTERNATE strategy with the wide angle reentrant corner problem.

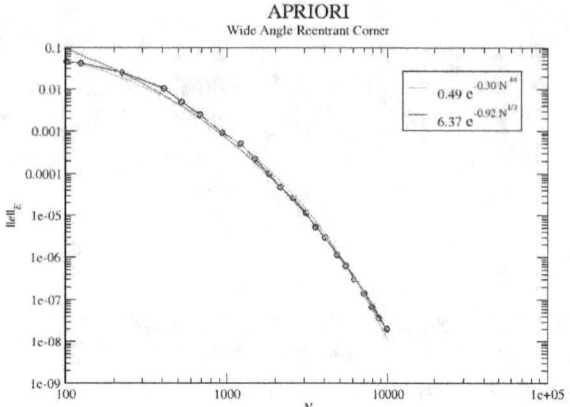

Figure 75: Log-Log plot of the convergence of the APRIORI strategy with the wide angle reentrant corner problem.

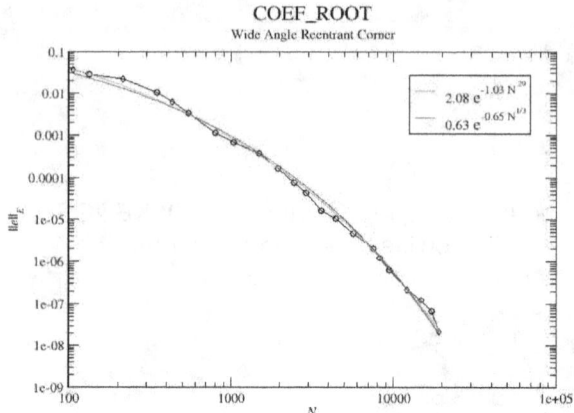

Figure 77: Log-Log plot of the convergence of the COEF-ROOT strategy with the wide angle reentrant corner problem.

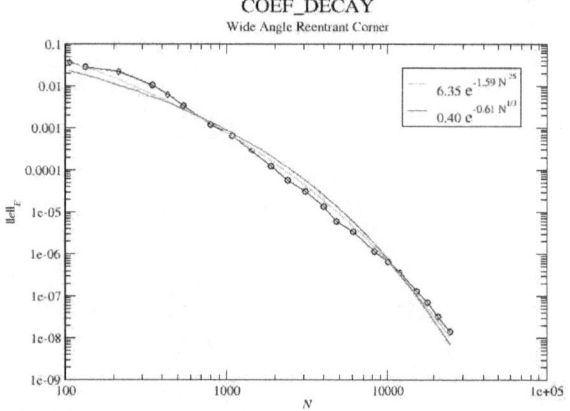

Figure 76: Log-Log plot of the convergence of the COEF-DECAY strategy with the wide angle reentrant corner problem.

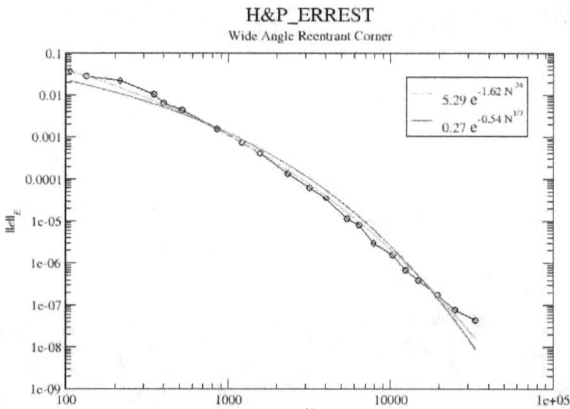

Figure 78: Log-Log plot of the convergence of the H&P-ERREST strategy with the wide angle reentrant corner problem.

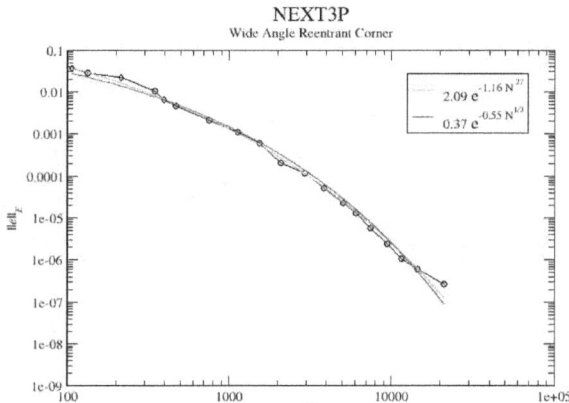

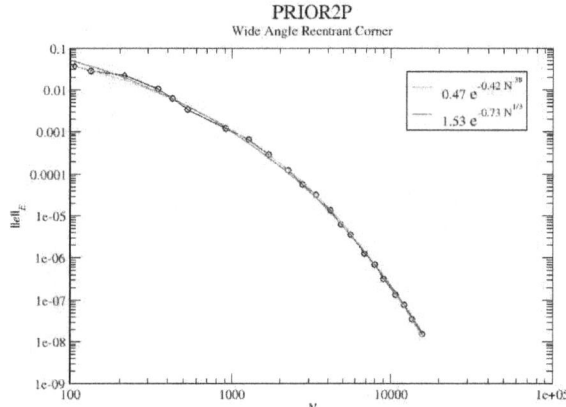

Figure 79: Log-Log plot of the convergence of the NEXT3P strategy with the wide angle reentrant corner problem.

Figure 81: Log-Log plot of the convergence of the PRIOR2P strategy with the wide angle reentrant corner problem.

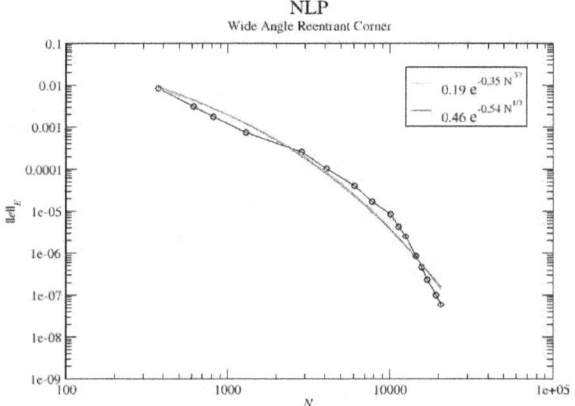

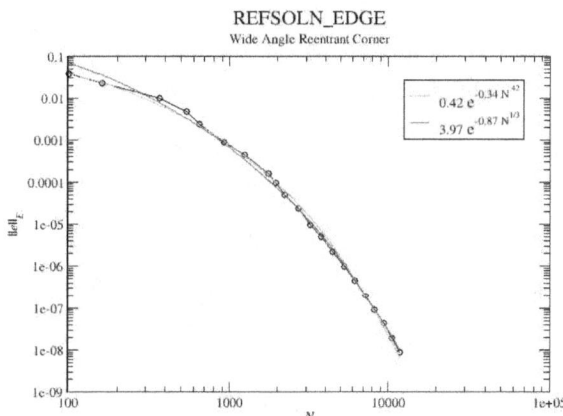

Figure 80: Log-Log plot of the convergence of the NLP strategy with the wide angle reentrant corner problem.

Figure 82: Log-Log plot of the convergence of the REFSOLN-EDGE strategy with the wide angle reentrant corner problem.

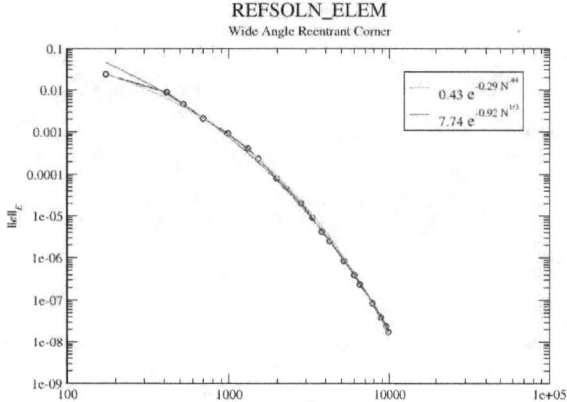

Figure 83: Log-Log plot of the convergence of the REFSOLN-ELEM strategy with the wide angle reentrant corner problem.

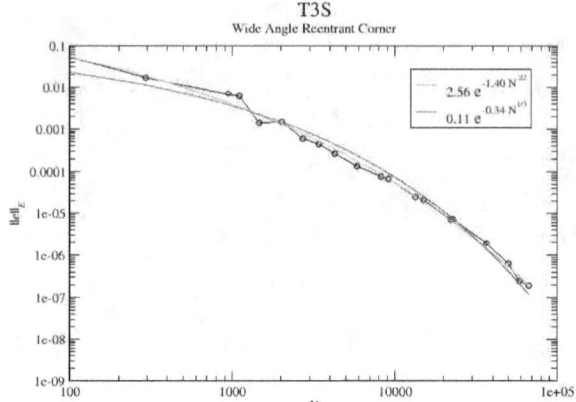

Figure 85: Log-Log plot of the convergence of the T3S strategy with the wide angle reentrant corner problem.

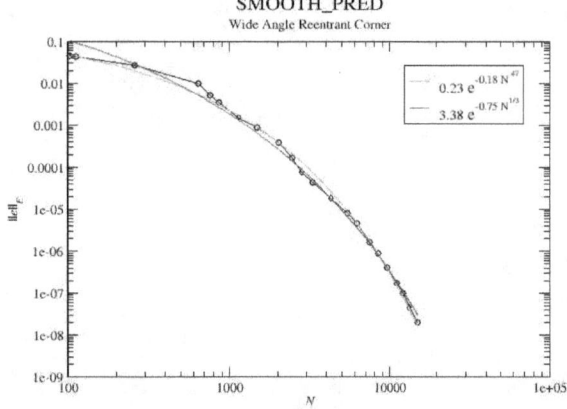

Figure 84: Log-Log plot of the convergence of the SMOOTH-PRED strategy with the wide angle reentrant corner problem.

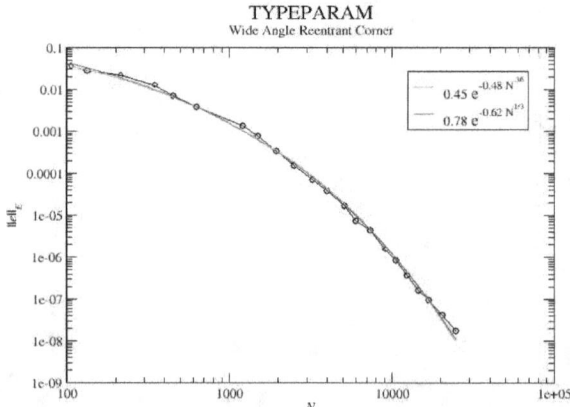

Figure 86: Log-Log plot of the convergence of the TYPEPARAM strategy with the wide angle reentrant corner problem.

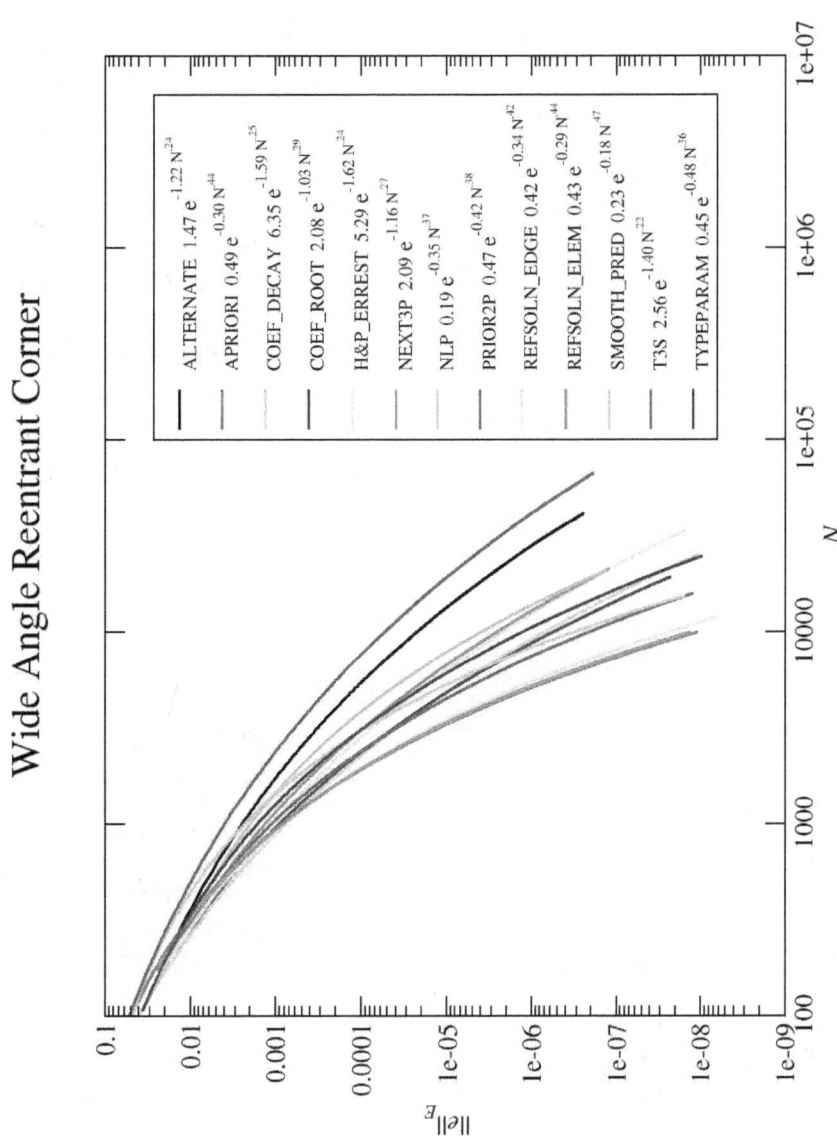

Figure 87: Log-Log plot of the convergence of all strategies with the wide angle reentrant corner problem.

strategy	A	B	C
ALTERNATE	1.47	-1.22	0.24
APRIORI	0.49	-0.30	0.44
COEFDECAY	6.35	-1.59	0.25
COEFROOT	2.08	-1.03	0.29
H&PERREST	5.29	-1.62	0.24
NEXT3P	2.09	-1.16	0.27
NLP	0.19	-0.35	0.37
PRIOR2P	0.47	-0.42	0.38
REFSOLNEDGE	0.42	-0.34	0.42
REFSOLNELEM	0.43	-0.29	0.44
SMOOTHPRED	0.23	-0.18	0.47
T3S	2.56	-1.40	0.22
TYPEPARAM	0.45	-0.48	0.36

Table 9: Parameters of the least squares fit for $\|e_{hp}\|_E = Ae^{BN_{dof}^C}$ for the wide angle reentrant corner problem.

strategy	A	B
ALTERNATE	0.13	-0.39
APRIORI	6.37	-0.92
COEFDECAY	0.40	-0.61
COEFROOT	0.63	-0.65
H&PERREST	0.27	-0.54
NEXT3P	0.37	-0.55
NLP	0.46	-0.54
PRIOR2P	1.53	-0.73
REFSOLNEDGE	3.97	-0.87
REFSOLNELEM	7.74	-0.92
SMOOTHPRED	3.38	-0.75
T3S	0.11	-0.34
TYPEPARAM	0.78	-0.62

Table 11: Parameters of the least squares fit for $\|e_{hp}\|_E = Ae^{BN_{dof}^{1/3}}$ for the wide angle reentrant corner problem.

strategy	factor
COEFDECAY	1.00
COEFROOT	1.05
H&PERREST	1.07
REFSOLNEDGE	1.10
NEXT3P	1.14
PRIOR2P	1.16
NLP	1.18
REFSOLNELEM	1.20
APRIORI	1.24
TYPEPARAM	1.27
ALTERNATE	1.35
SMOOTHPRED	1.66
T3S	1.88

Table 10: Factor by which N is larger than the best strategy for the wide angle reentrant corner problem at low accuracy, 1.0×10^{-2}.

strategy	factor
APRIORI	1.00
REFSOLNELEM	1.02
REFSOLNEDGE	1.07
PRIOR2P	1.46
SMOOTHPRED	1.62
COEFROOT	1.71
COEFDECAY	1.79
TYPEPARAM	2.06
H&PERREST	2.43
NEXT3P	2.52
NLP	2.70
ALTERNATE	5.49
T3S	8.11

Table 12: Factor by which N is larger than the best strategy for the wide angle reentrant corner problem at high accuracy, 1.0×10^{-6}.

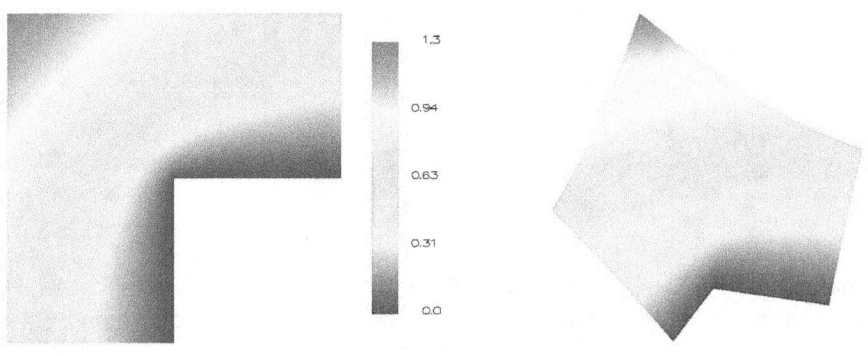

Figure 88: The solution of the L-shaped domain problem.

5.4 Reentrant Corner, L-Shaped Domain

The reentrant corner problem (Section 5.2) with $\omega=3\pi/2$ is the classic "L domain" problem which is used as an example in many papers on adaptive grid refinement. $\tau=10^{-4}$ for the grid images. The APRIORI strategy refines by h if the element contains the origin and by p otherwise. For this problem, the cube root vs. log convergence plots are shown in addition to the log-log plots.

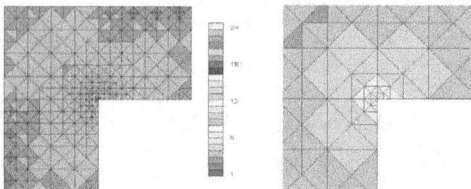

Figure 89: Example grid for the ALTERNATE strategy with the L-shaped domain problem, including details at the singularity.

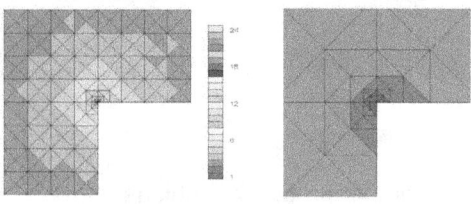

Figure 90: Example grid for the APRIORI strategy with the L-shaped domain problem, including details at the singularity.

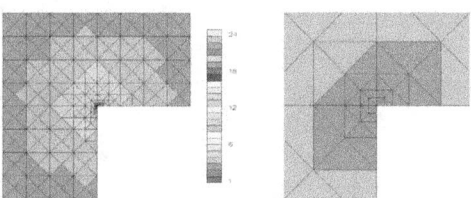

Figure 91: Example grid for the COEFDECAY strategy with the L-shaped domain problem, including details at the singularity.

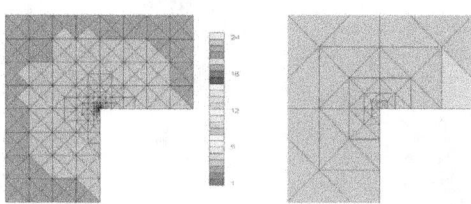

Figure 92: Example grid for the COEFROOT strategy with the L-shaped domain problem, including details at the singularity.

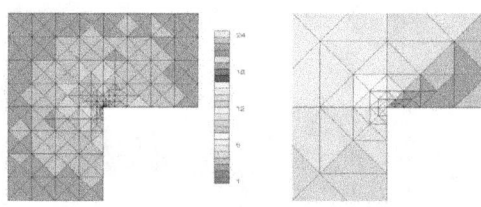

Figure 93: Example grid for the H&PERREST strategy with the L-shaped domain problem, including details at the singularity.

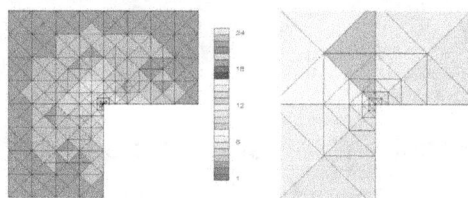

Figure 94: Example grid for the NEXT3P strategy with the L-shaped domain problem, including details at the singularity.

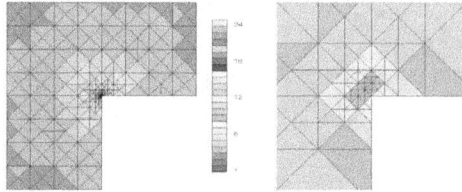

Figure 95: Example grid for the NLP strategy with the L-shaped domain problem, including details at the singularity.

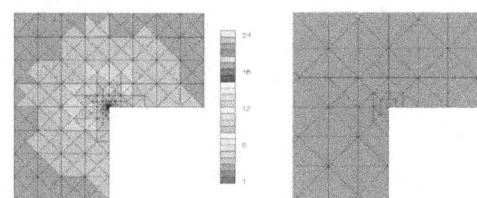

Figure 96: Example grid for the PRIOR2P strategy with the L-shaped domain problem, including details at the singularity.

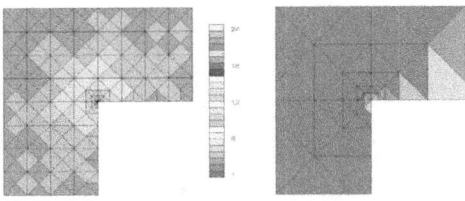

Figure 97: Example grid for the REFSOLNEDGE strategy with the L-shaped domain problem, including details at the singularity.

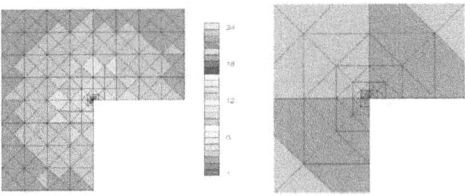

Figure 98: Example grid for the REFSOLNELEM strategy with the L-shaped domain problem, including details at the singularity.

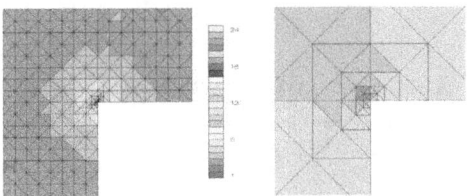

Figure 99: Example grid for the SMOOTHPRED strategy with the L-shaped domain problem, including details at the singularity.

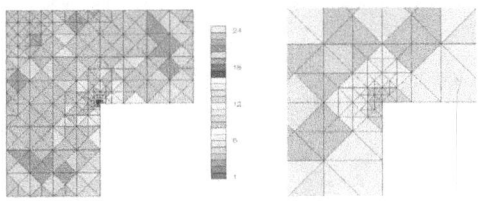

Figure 100: Example grid for the T3S strategy with the L-shaped domain problem, including details at the singularity.

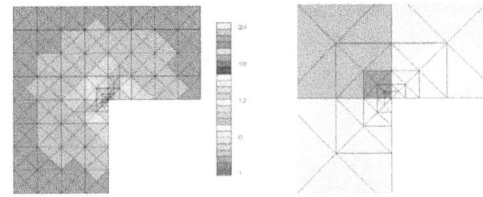

Figure 101: Example grid for the TYPEPARAM strategy with the L-shaped domain problem, including details at the singularity.

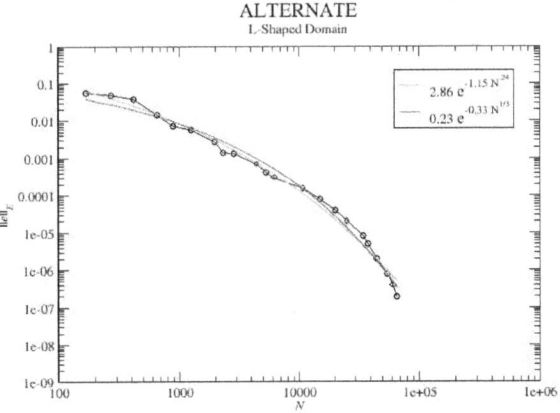

Figure 102: Log-Log plot of the convergence of the ALTERNATE strategy with the L-shaped domain problem.

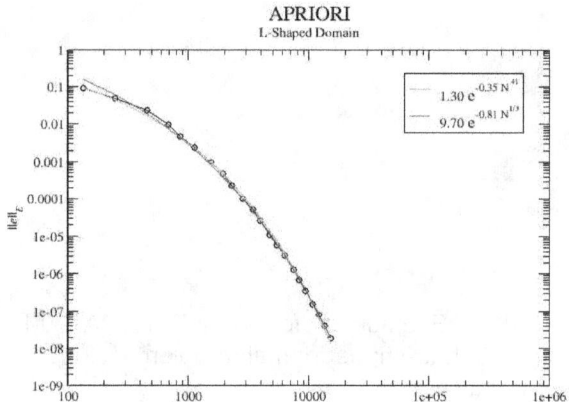

Figure 103: Log-Log plot of the convergence of the APRIORI strategy with the L-shaped domain problem.

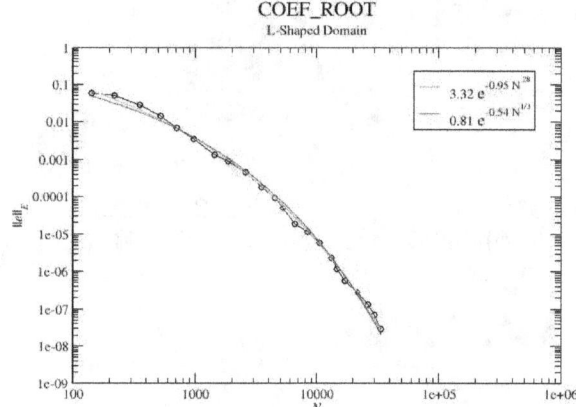

Figure 105: Log-Log plot of the convergence of the COEF-ROOT strategy with the L-shaped domain problem.

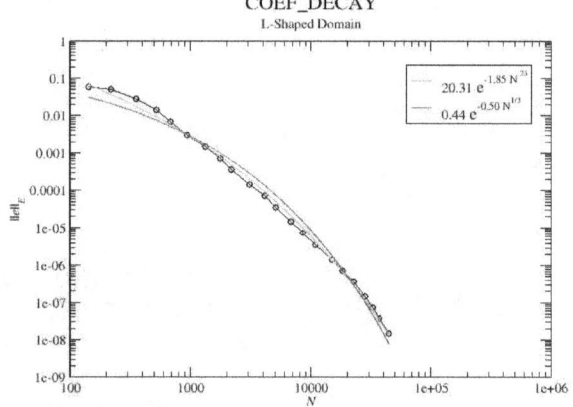

Figure 104: Log-Log plot of the convergence of the COEF-DECAY strategy with the L-shaped domain problem.

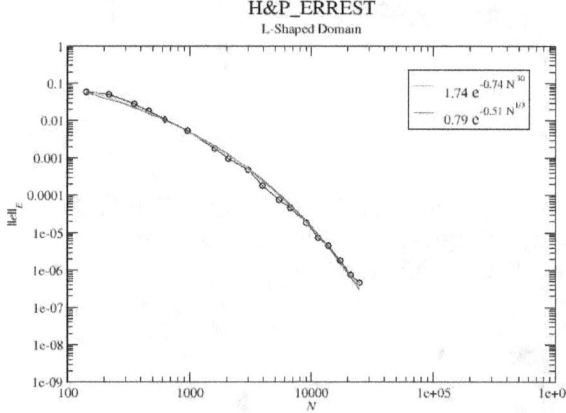

Figure 106: Log-Log plot of the convergence of the H&P-ERREST strategy with the L-shaped domain problem.

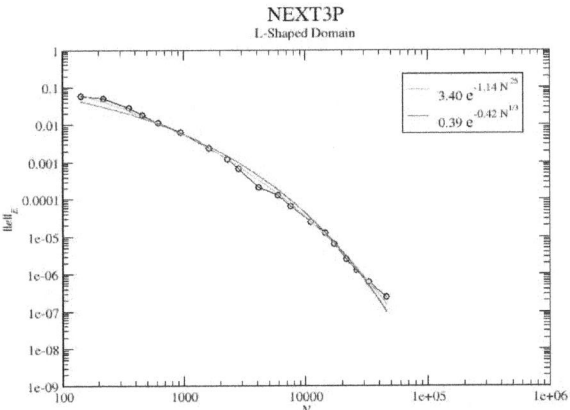

Figure 107: Log-Log plot of the convergence of the NEXT3P strategy with the L-shaped domain problem.

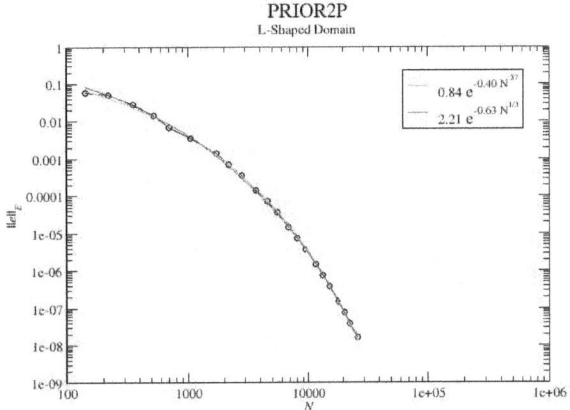

Figure 109: Log-Log plot of the convergence of the PRIOR2P strategy with the L-shaped domain problem.

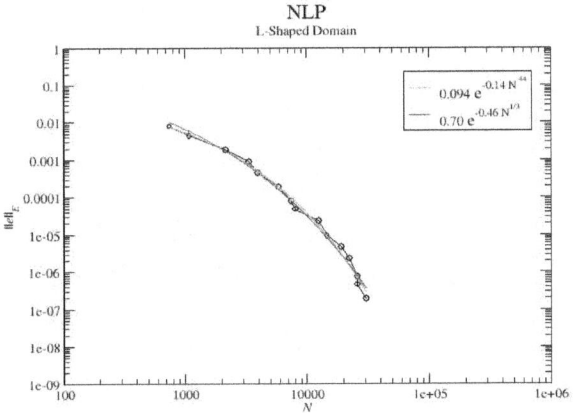

Figure 108: Log-Log plot of the convergence of the NLP strategy with the L-shaped domain problem.

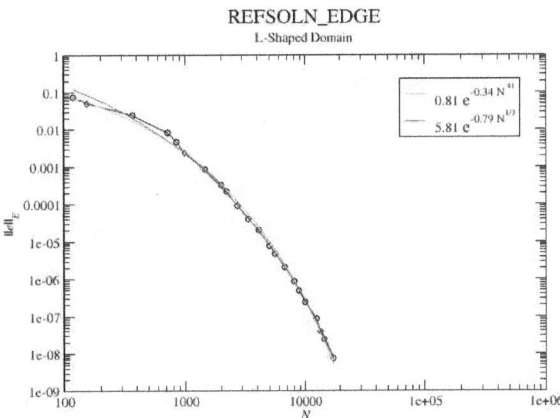

Figure 110: Log-Log plot of the convergence of the REFSOLN-EDGE strategy with the L-shaped domain problem.

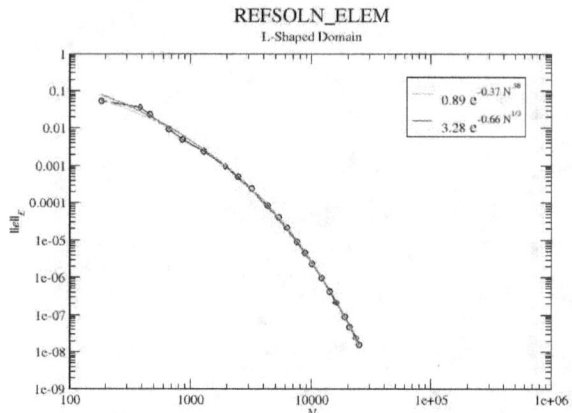

Figure 111: Log-Log plot of the convergence of the REFSOLN-ELEM strategy with the L-shaped domain problem.

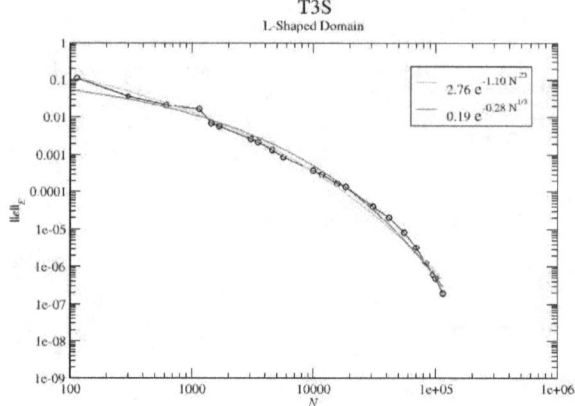

Figure 113: Log-Log plot of the convergence of the T3S strategy with the L-shaped domain problem.

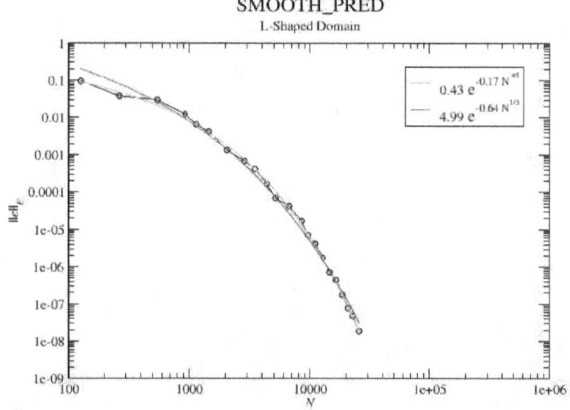

Figure 112: Log-Log plot of the convergence of the SMOOTH-PRED strategy with the L-shaped domain problem.

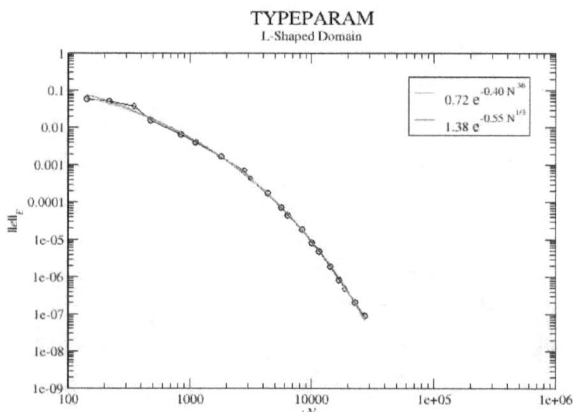

Figure 114: Log-Log plot of the convergence of the TYPEPARAM strategy with the L-shaped domain problem.

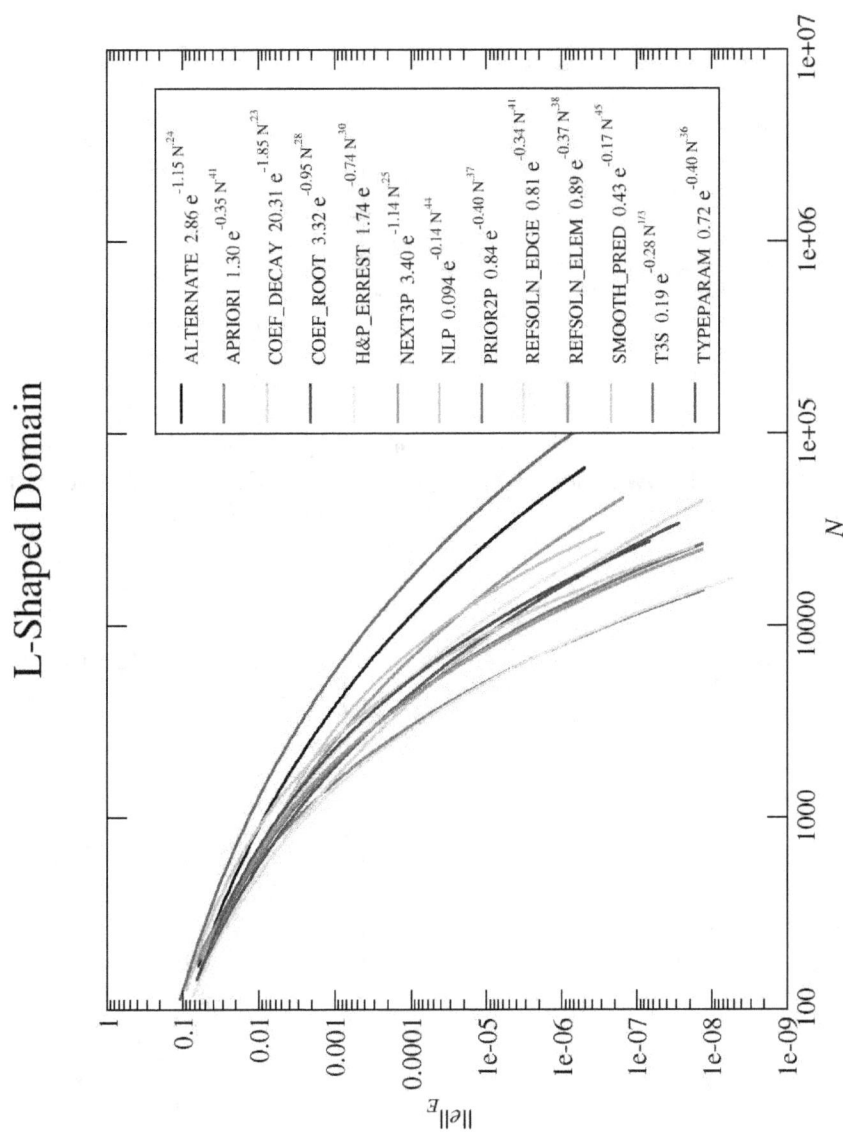

Figure 115: Log-Log plot of the convergence of all strategies with the L-shaped domain problem.

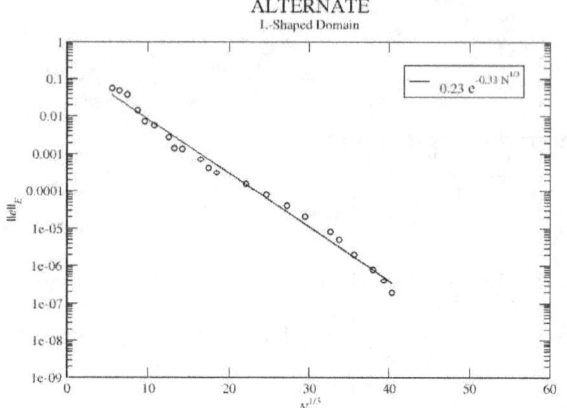
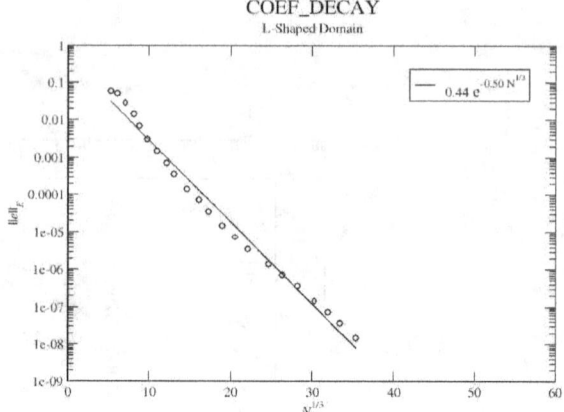

Figure 116: Cuberoot vs. Logplot of the convergence of the ALTERNATE strategy with the L-shaped domain problem.

Figure 118: Cuberoot vs. Logplot of the convergence of the COEF-DECAY strategy with the L-shaped domain problem.

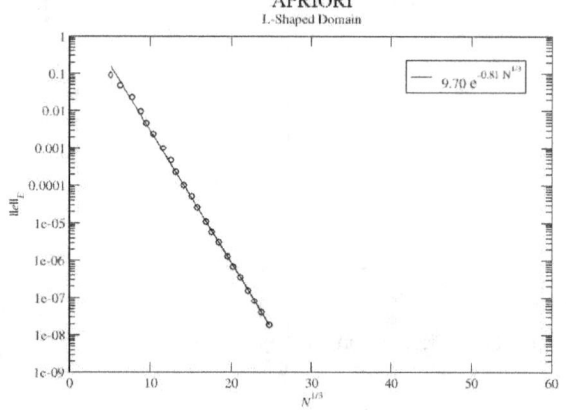
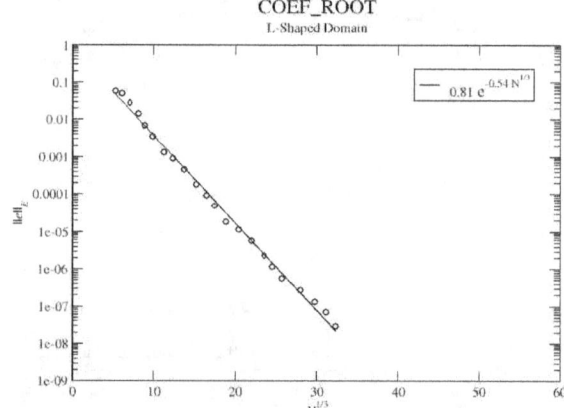

Figure 117: Cuberoot vs. Logplot of the convergence of the APRIORI strategy with the L-shaped domain problem.

Figure 119: Cuberoot vs. Logplot of the convergence of the COEF-ROOT strategy with the L-shaped domain problem.

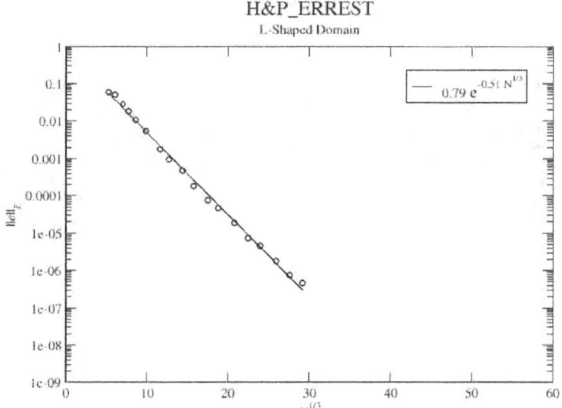
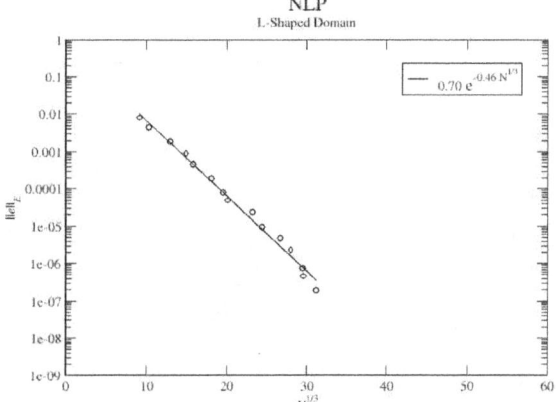

Figure 120: Cuberoot vs. Logplot of the convergence of the H&P-ERREST strategy with the L-shaped domain problem.

Figure 122: Cuberoot vs. Logplot of the convergence of the NLP strategy with the L-shaped domain problem.

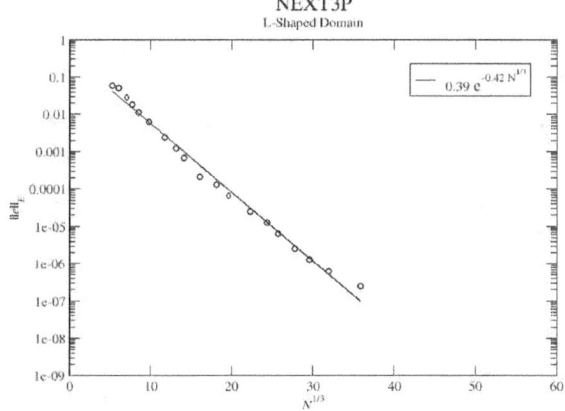
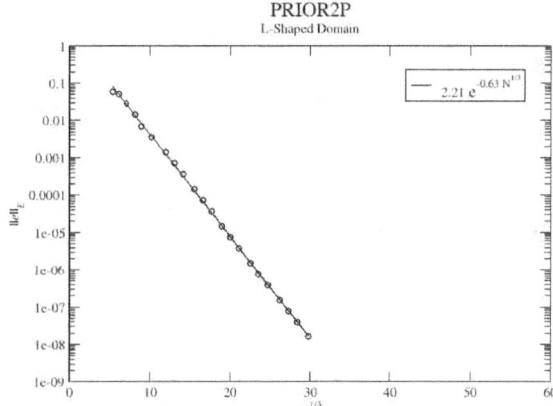

Figure 121: Cuberoot vs. Logplot of the convergence of the NEXT3P strategy with the L-shaped domain problem.

Figure 123: Cuberoot vs. Logplot of the convergence of the PRIOR2P strategy with the L-shaped domain problem.

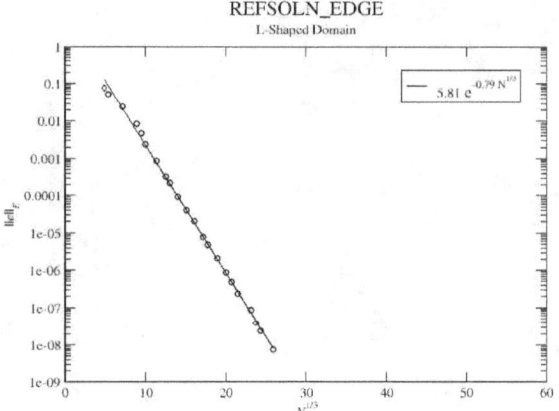

Figure 124: Cube root vs. Log plot of the convergence of the REFSOLN_EDGE strategy with the L-shaped domain problem.

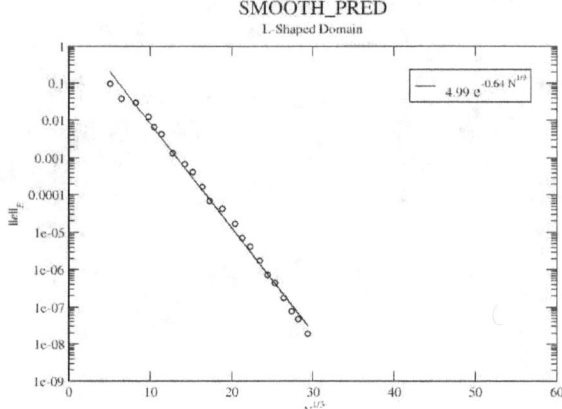

Figure 126: Cube root vs. Log plot of the convergence of the SMOOTH_PRED strategy with the L-shaped domain problem.

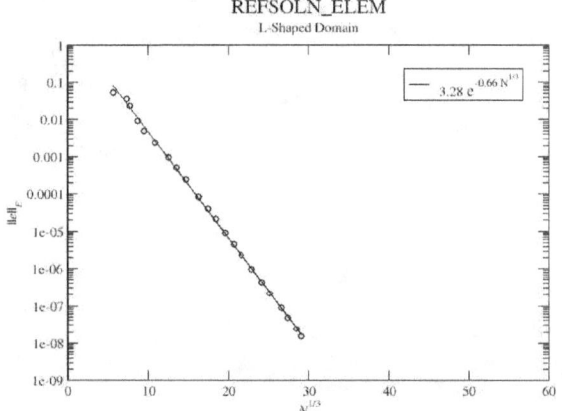

Figure 125: Cube root vs. Log plot of the convergence of the REFSOLN_ELEM strategy with the L-shaped domain problem.

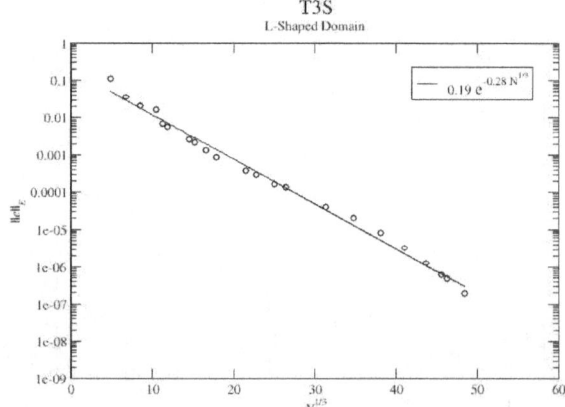

Figure 127: Cube root vs. Log plot of the convergence of the T3S strategy with the L-shaped domain problem.

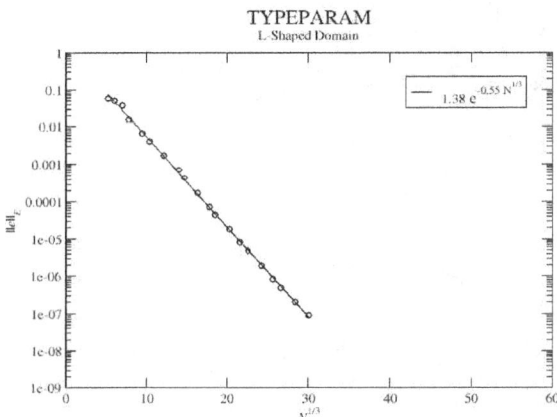

Figure 128: Cuberoot vs. Logplot of the convergence of the TYPEPARAM strategy with the L-shaped domain problem.

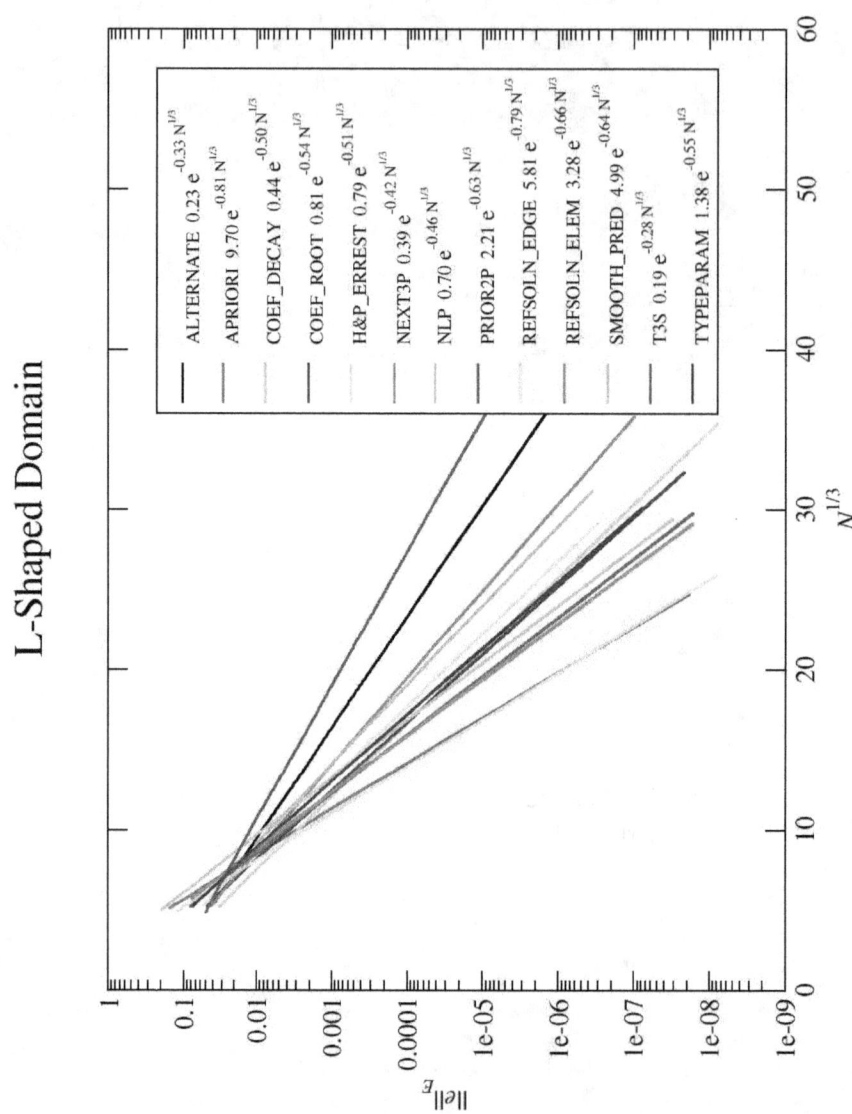

Figure 129: Cube root vs. Log plot of the convergence of all strategies with the L-shaped domain problem.

strategy	A	B	C
ALTERNATE	2.86	-1.15	0.24
APRIORI	1.30	-0.35	0.41
COEFDECAY	20.31	-1.85	0.23
COEFROOT	3.32	-0.95	0.28
H&PERREST	1.74	-0.74	0.30
NEXT3P	3.40	-1.14	0.25
NLP	0.094	-0.14	0.44
PRIOR2P	0.84	-0.40	0.37
REFSOLNEDGE	0.81	-0.34	0.41
REFSOLNELEM	0.89	-0.37	0.38
SMOOTHPRED	0.43	-0.17	0.45
T3S	2.76	-1.10	0.23
TYPEPARAM	0.72	-0.40	0.36

Table 13: Parameters of the least squares fit for $\|e_{hp}\|_E = A e^{B N_{dof}^C}$ for the L-shaped domain problem.

strategy	A	B
ALTERNATE	0.23	-0.33
APRIORI	9.70	-0.81
COEFDECAY	0.44	-0.50
COEFROOT	0.81	-0.54
H&PERREST	0.79	-0.51
NEXT3P	0.39	-0.42
NLP	0.70	-0.46
PRIOR2P	2.21	-0.63
REFSOLNEDGE	5.81	-0.79
REFSOLNELEM	3.28	-0.66
SMOOTHPRED	4.99	-0.64
T3S	0.19	-0.28
TYPEPARAM	1.38	-0.55

Table 15: Parameters of the least squares fit for $\|e_{hp}\|_E = A e^{B N_{dof}^{1/3}}$ for the L-shaped domain problem.

strategy	factor
COEFDECAY	1.00
REFSOLNEDGE	1.01
COEFROOT	1.13
NLP	1.14
APRIORI	1.18
PRIOR2P	1.24
H&PERREST	1.27
NEXT3P	1.32
REFSOLNELEM	1.32
TYPEPARAM	1.39
ALTERNATE	1.77
SMOOTHPRED	1.77
T3S	2.54

Table 14: Factor by which N is larger than the best strategy for the L-shaped domain problem at low accuracy, 1.0×10^{-2}.

strategy	factor
APRIORI	1.00
REFSOLNEDGE	1.00
REFSOLNELEM	1.54
PRIOR2P	1.61
SMOOTHPRED	1.77
COEFROOT	2.03
COEFDECAY	2.08
TYPEPARAM	2.09
H&PERREST	2.48
NLP	3.03
NEXT3P	3.67
ALTERNATE	6.90
T3S	11.55

Table 16: Factor by which N is larger than the best strategy for the L-shaped domain problem at high accuracy, 1.0×10^{-6}.

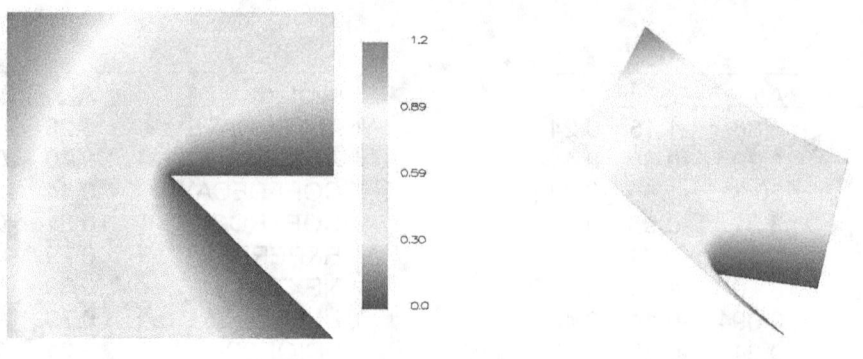

Figure 130: The solution of the narrow angle reentrant corner problem.

5.5 Reentrant Corner, Narrow Angle

This is the reentrant corner problem (Section 5.2) with $\omega = 7\pi/4$. $\tau = 10^{-4}$ for the grid images. The APRIORI strategy refines by h if the element contains the origin and by p otherwise.

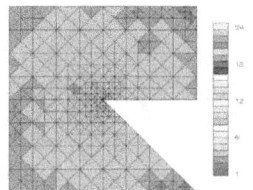

Figure 131: Example grid for the ALTERNATE strategy with the narrow angle reentrant corner problem, including details at the singularity.

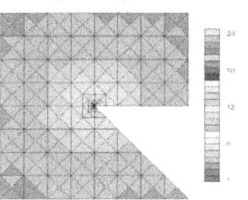
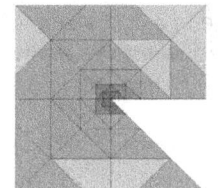

Figure 132: Example grid for the APRIORI strategy with the narrow angle reentrant corner problem, including details at the singularity.

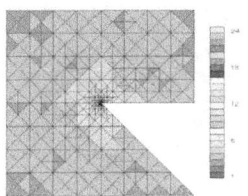
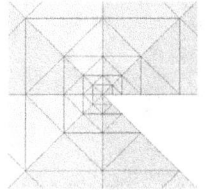

Figure 135: Example grid for the H&PERREST strategy with the narrow angle reentrant corner problem, including details at the singularity.

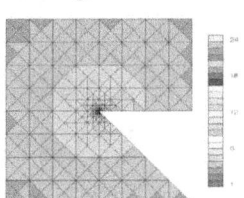
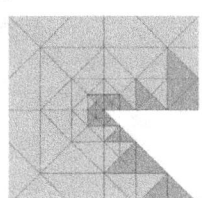

Figure 133: Example grid for the COEFDECAY strategy with the narrow angle reentrant corner problem, including details at the singularity.

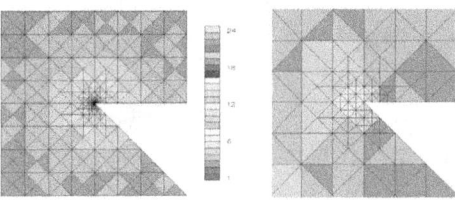

Figure 136: Example grid for the NEXT3P strategy with the narrow angle reentrant corner problem, including details at the singularity.

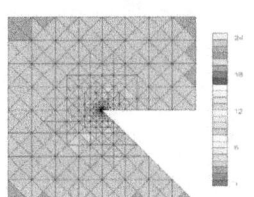
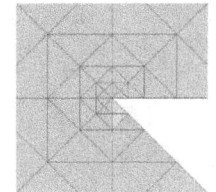

Figure 134: Example grid for the COEFROOT strategy with the narrow angle reentrant corner problem, including details at the singularity.

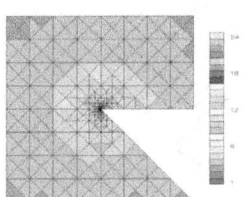
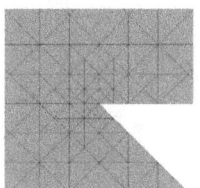

Figure 137: Example grid for the NLP strategy with the narrow angle reentrant corner problem, including details at the singularity.

Figure 138: Example grid for the PRIOR2P strategy with the narrow angle reentrant corner problem, including details at the singularity.

Figure 139: Example grid for the REFSOLNEDGE strategy with the narrow angle reentrant corner problem, including details at the singularity.

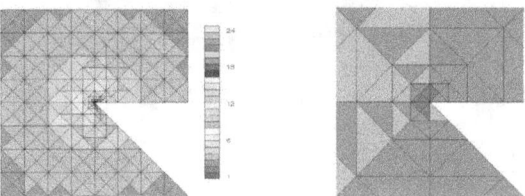

Figure 140: Example grid for the REFSOLNELEM strategy with the narrow angle reentrant corner problem, including details at the singularity.

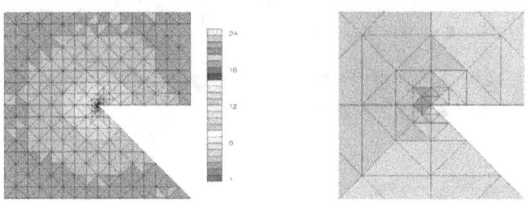

Figure 141: Example grid for the SMOOTHPRED strategy with the narrow angle reentrant corner problem, including details at the singularity.

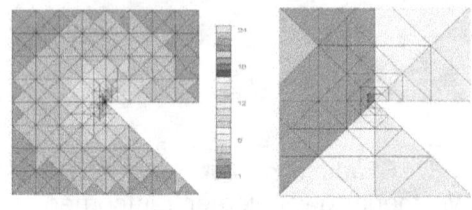

Figure 143: Example grid for the TYPEPARAM strategy with the narrow angle reentrant corner problem, including details at the singularity.

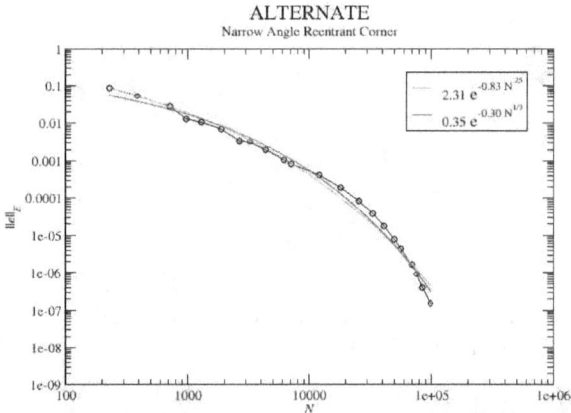

Figure 144: Log-Log plot of the convergence of the ALTERNATE strategy with the narrow angle reentrant corner problem.

Figure 142: Example grid for the T3S strategy with the narrow angle reentrant corner problem, including details at the singularity.

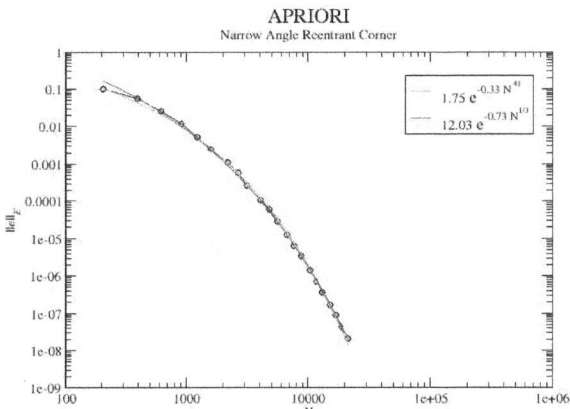

Figure 145: Log-Log plot of the convergence of the APRIORI strategy with the narrow angle reentrant corner problem.

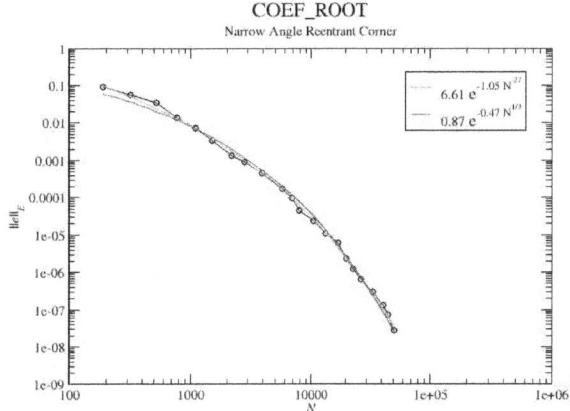

Figure 147: Log-Log plot of the convergence of the COEF-ROOT strategy with the narrow angle reentrant corner problem.

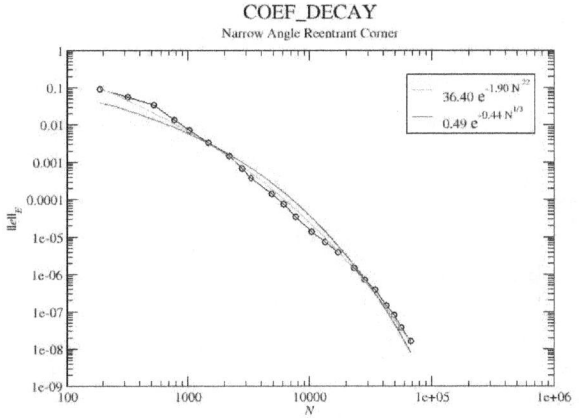

Figure 146: Log-Log plot of the convergence of the COEF-DECAY strategy with the narrow angle reentrant corner problem.

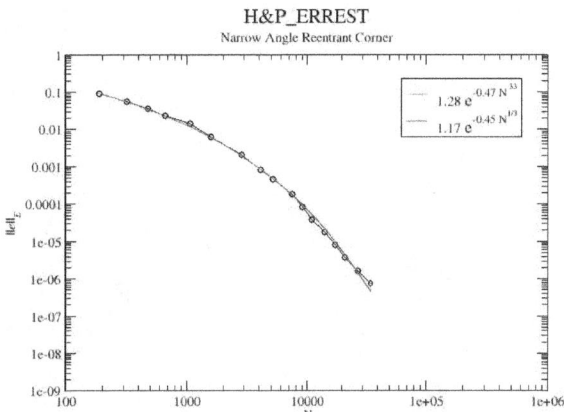

Figure 148: Log-Log plot of the convergence of the H&P-ERREST strategy with the narrow angle reentrant corner problem.

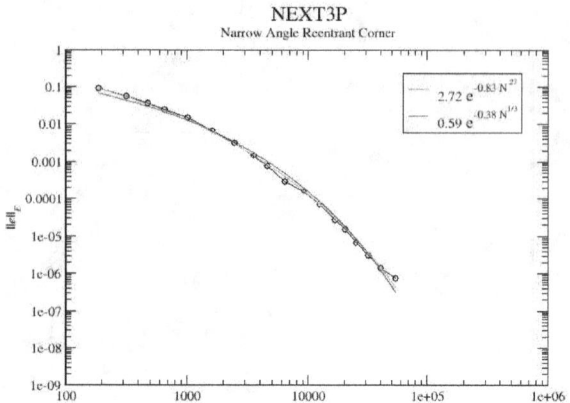

Figure 149: Log-Log plot of the convergence of the NEXT3P strategy with the narrow angle reentrant corner problem.

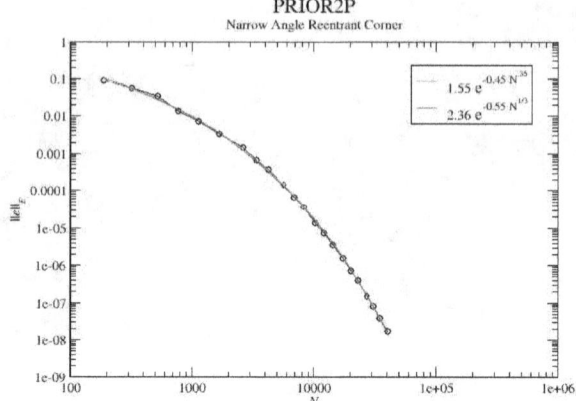

Figure 151: Log-Log plot of the convergence of the PRIOR2P strategy with the narrow angle reentrant corner problem.

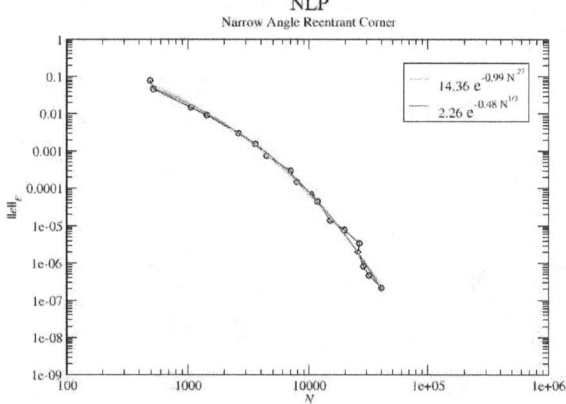

Figure 150: Log-Log plot of the convergence of the NLP strategy with the narrow angle reentrant corner problem.

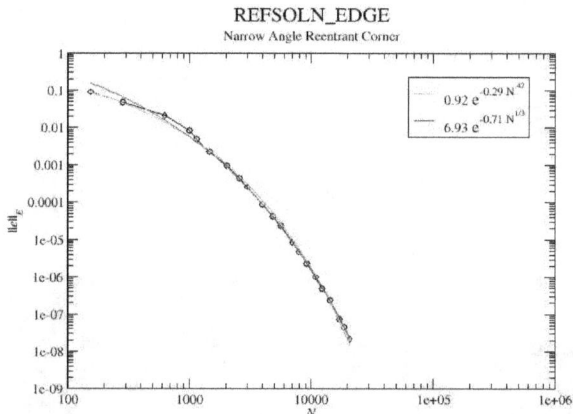

Figure 152: Log-Log plot of the convergence of the REFSOLN-EDGE strategy with the narrow angle reentrant corner problem.

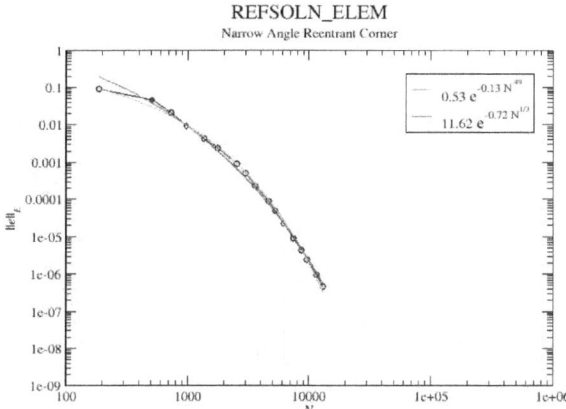

Figure 153: Log-Log plot of the convergence of the REFSOLN-ELEM strategy with the narrow angle reentrant corner problem.

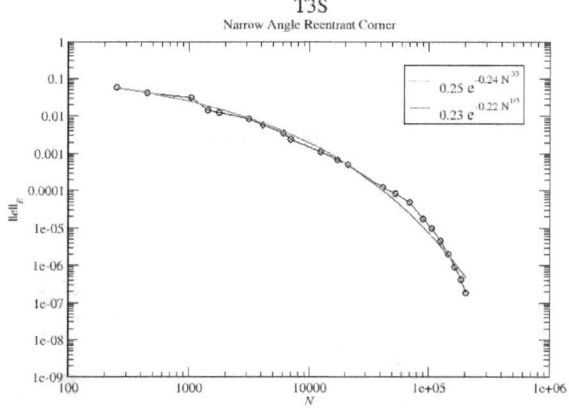

Figure 155: Log-Log plot of the convergence of the T3S strategy with the narrow angle reentrant corner problem.

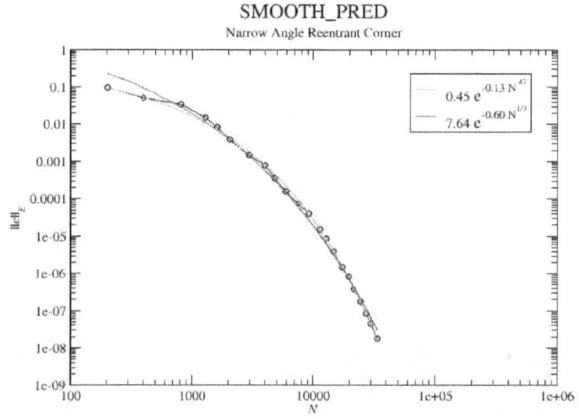

Figure 154: Log-Log plot of the convergence of the SMOOTH-PRED strategy with the narrow angle reentrant corner problem.

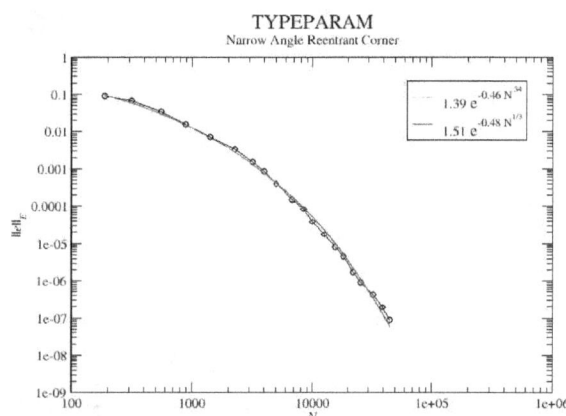

Figure 156: Log-Log plot of the convergence of the TYPEPARAM strategy with the narrow angle reentrant corner problem.

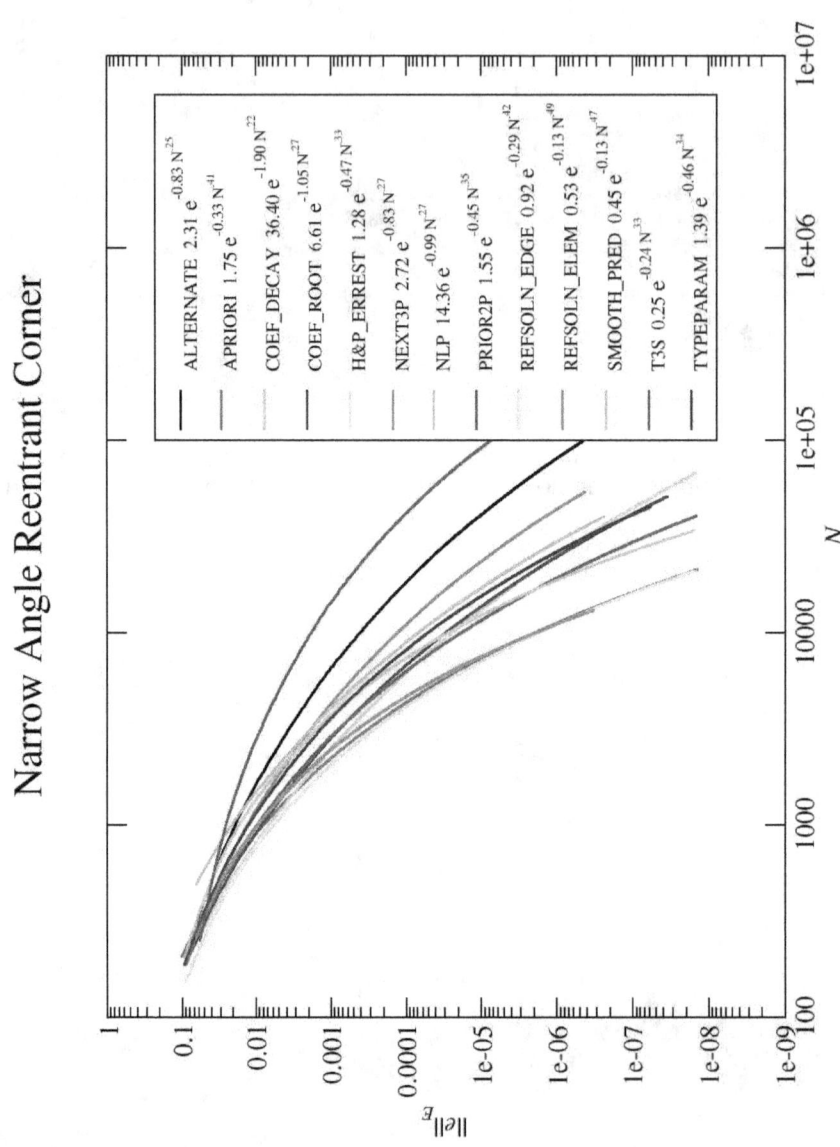

Figure 157: Log-Log plot of the convergence of all strategies with the narrow angle reentrant corner problem.

strategy	A	B	C
ALTERNATE	2.31	-0.83	0.25
APRIORI	1.75	-0.33	0.41
COEFDECAY	36.40	-1.90	0.22
COEFROOT	6.61	-1.05	0.27
H&PERREST	1.28	-0.47	0.33
NEXT3P	2.72	-0.83	0.27
NLP	14.36	-0.99	0.27
PRIOR2P	1.55	-0.45	0.35
REFSOLNEDGE	0.92	-0.29	0.42
REFSOLNELEM	0.53	-0.13	0.49
SMOOTHPRED	0.45	-0.13	0.47
T3S	0.25	-0.24	0.33
TYPEPARAM	1.39	-0.46	0.34

Table 17: Parameters of the least squares fit for $\|e_{hp}\|_E = A e^{BN_{dof}^C}$ for the narrow angle reentrant corner problem.

strategy	A	B
ALTERNATE	0.35	-0.30
APRIORI	12.03	-0.73
COEFDECAY	0.49	-0.44
COEFROOT	0.87	-0.47
H&PERREST	1.17	-0.45
NEXT3P	0.59	-0.38
NLP	2.26	-0.48
PRIOR2P	2.36	-0.55
REFSOLNEDGE	6.93	-0.71
REFSOLNELEM	11.62	-0.72
SMOOTHPRED	7.64	-0.60
T3S	0.23	-0.22
TYPEPARAM	1.51	-0.48

Table 19: Parameters of the least squares fit for $\|e_{hp}\|_E = A e^{BN_{dof}^{1/3}}$ for the narrow angle reentrant corner problem.

strategy	factor
REFSOLNEDGE	1.00
COEFDECAY	1.06
APRIORI	1.18
COEFROOT	1.19
REFSOLNELEM	1.28
PRIOR2P	1.29
TYPEPARAM	1.51
H&PERREST	1.54
NEXT3P	1.57
SMOOTHPRED	1.78
NLP	1.95
ALTERNATE	2.14
T3S	3.66

Table 18: Factor by which N is larger than the best strategy for the narrow angle reentrant corner problem at low accuracy, 1.0×10^{-2}.

strategy	factor
REFSOLNEDGE	1.00
REFSOLNELEM	1.00
APRIORI	1.01
SMOOTHPRED	1.69
PRIOR2P	1.74
COEFROOT	2.25
COEFDECAY	2.28
TYPEPARAM	2.38
NLP	2.68
H&PERREST	2.69
NEXT3P	3.97
ALTERNATE	7.30
T3S	15.52

Table 20: Factor by which N is larger than the best strategy for the narrow angle reentrant corner problem at high accuracy, 1.0×10^{-6}.

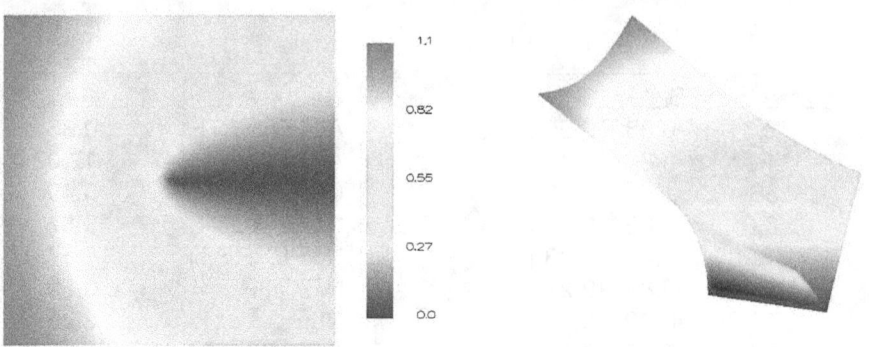

Figure 158: The solution of the slit domain problem.

5.6 Reentrant Corner, Slit

This is the reentrant corner problem (Section 5.2) with $\omega = 2\pi$. This results in a domain that has a slit along the positive x axis. $\tau = 10^{-4}$ for the grid images. The APRIORI strategy refines by h if the element contains the origin and by p otherwise.

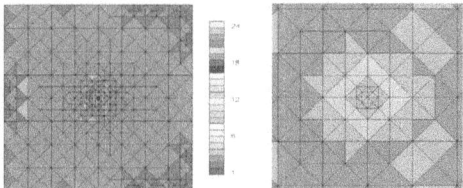

Figure 159: Example grid for the ALTERNATE strategy with the slit domain problem, including details at the singularity.

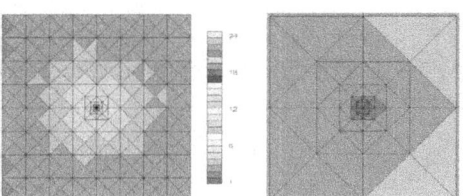

Figure 160: Example grid for the APRIORI strategy with the slit domain problem, including details at the singularity.

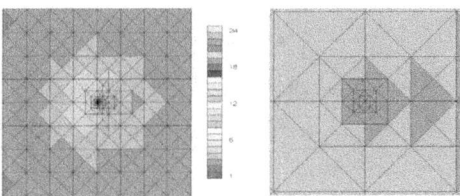

Figure 161: Example grid for the COEFDECAY strategy with the slit domain problem, including details at the singularity.

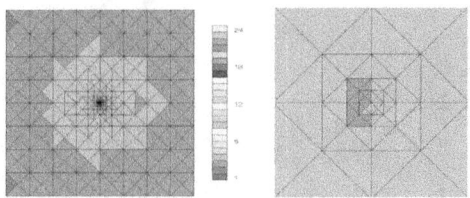

Figure 162: Example grid for the COEFROOT strategy with the slit domain problem, including details at the singularity.

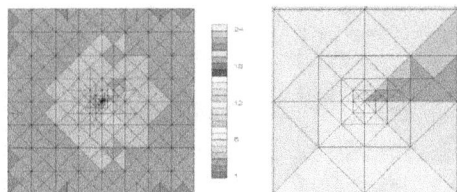

Figure 163: Example grid for the H&PERREST strategy with the slit domain problem, including details at the singularity.

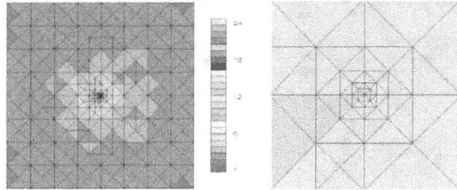

Figure 164: Example grid for the NEXT3P strategy with the slit domain problem, including details at the singularity.

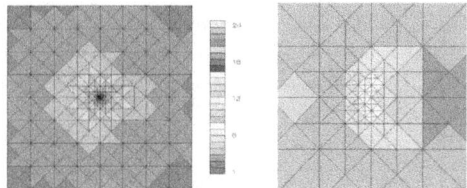

Figure 165: Example grid for the NLP strategy with the slit domain problem, including details at the singularity.

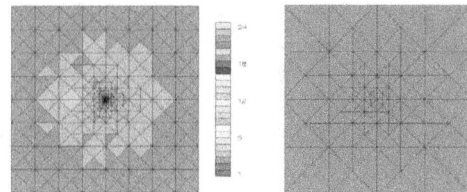

Figure 166: Example grid for the PRIOR2P strategy with the slit domain problem, including details at the singularity.

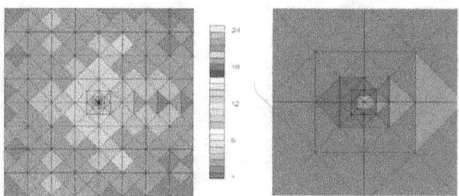

Figure 167: Example grid for the REFSOLNEDGE strategy with the slit domain problem, including details at the singularity.

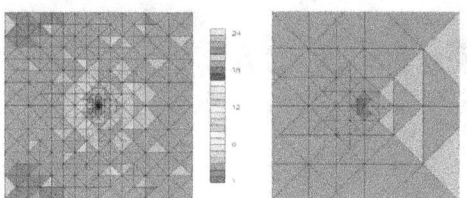

Figure 168: Example grid for the REFSOLNELEM strategy with the slit domain problem, including details at the singularity.

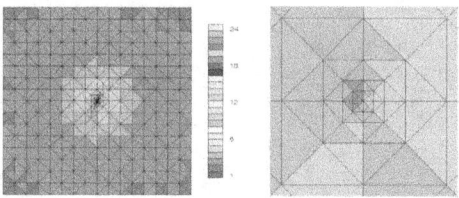

Figure 169: Example grid for the SMOOTHPRED strategy with the slit domain problem, including details at the singularity.

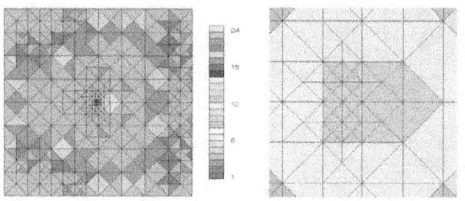

Figure 170: Example grid for the T3S strategy with the slit domain problem, including details at the singularity.

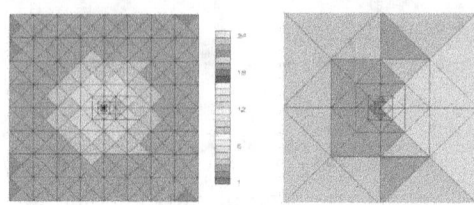

Figure 171: Example grid for the TYPEPARAM strategy with the slit domain problem, including details at the singularity.

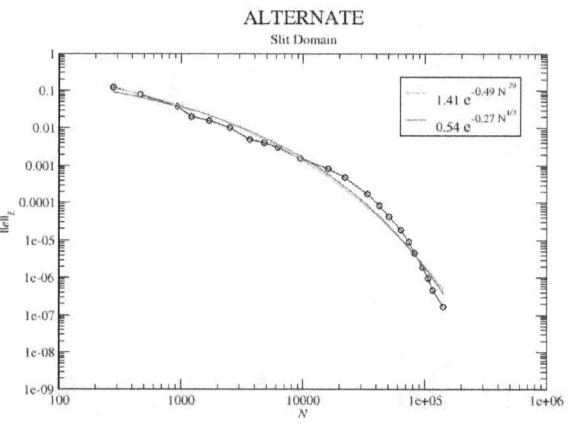

Figure 172: Log-Log plot of the convergence of the ALTERNATE strategy with the slit domain problem.

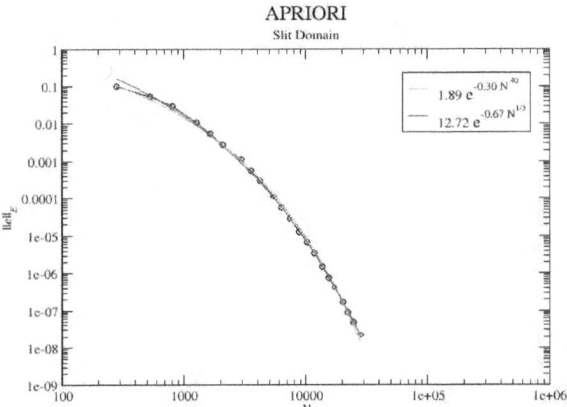

Figure 173: Log-Log plot of the convergence of the APRIORI strategy with the slit domain problem.

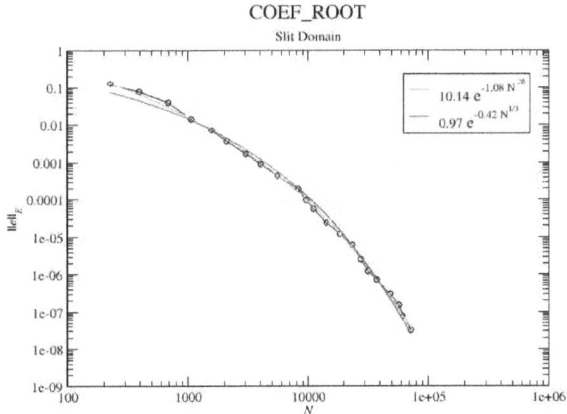

Figure 175: Log-Log plot of the convergence of the COEF-ROOT strategy with the slit domain problem.

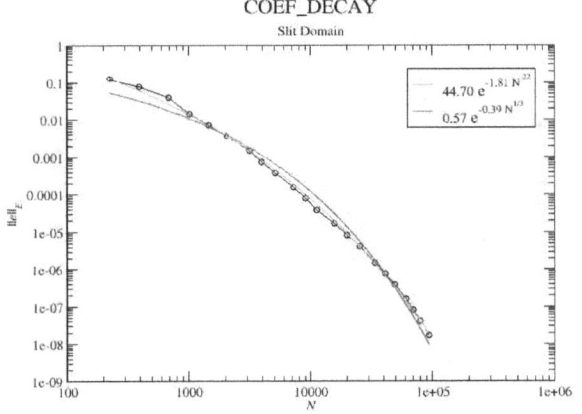

Figure 174: Log-Log plot of the convergence of the COEF-DECAY strategy with the slit domain problem.

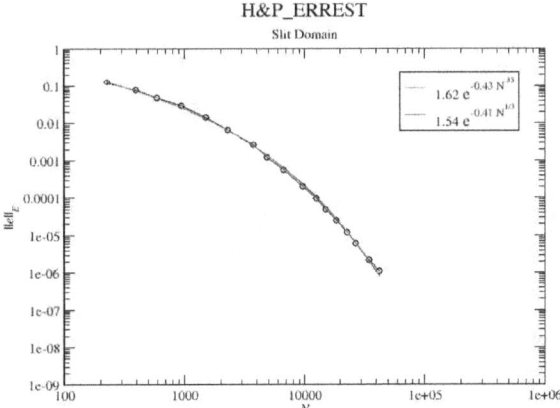

Figure 176: Log-Log plot of the convergence of the H&P-ERREST strategy with the slit domain problem.

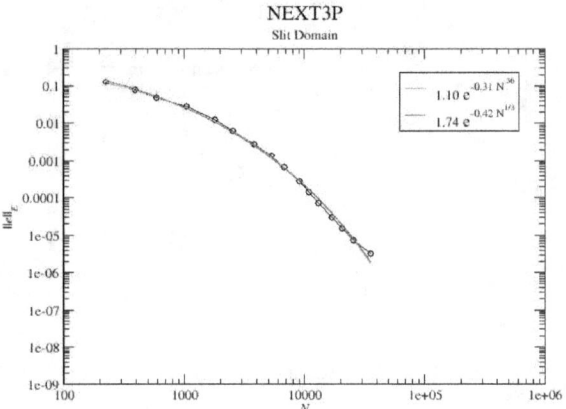

Figure 177: Log-Log plot of the convergence of the NEXT3P strategy with the slit domain problem.

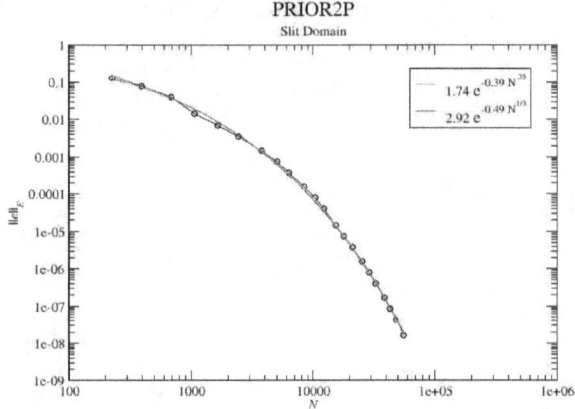

Figure 179: Log-Log plot of the convergence of the PRIOR2P strategy with the slit domain problem.

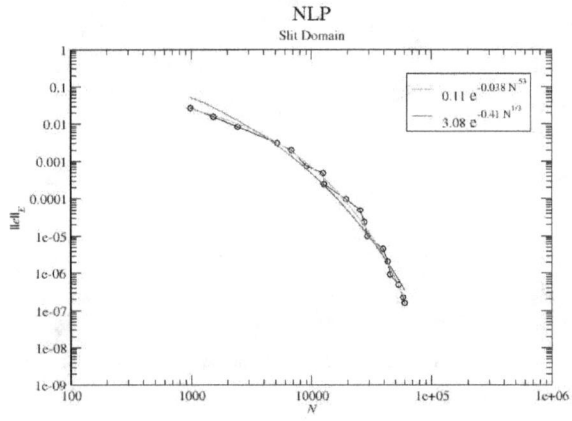

Figure 178: Log-Log plot of the convergence of the NLP strategy with the slit domain problem.

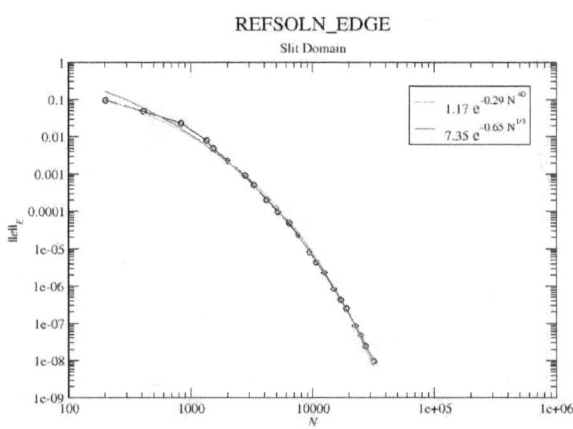

Figure 180: Log-Log plot of the convergence of the REFSOLN-EDGE strategy with the slit domain problem.

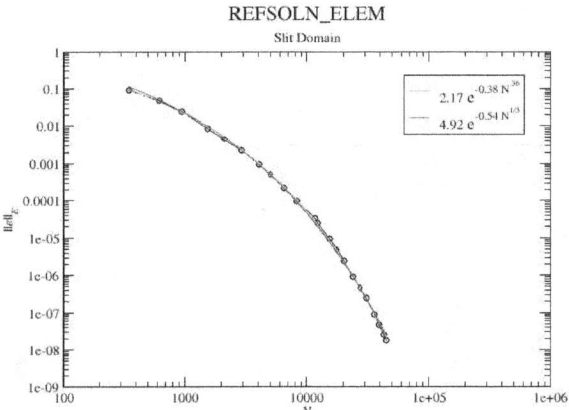

Figure 181: Log-Log plot of the convergence of the REFSOLN-ELEM strategy with the slit domain problem.

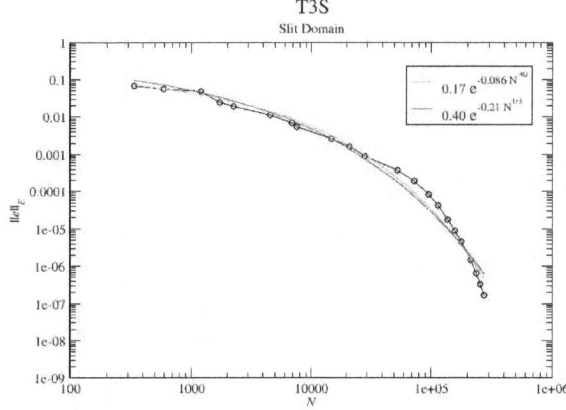

Figure 183: Log-Log plot of the convergence of the T3S strategy with the slit domain problem.

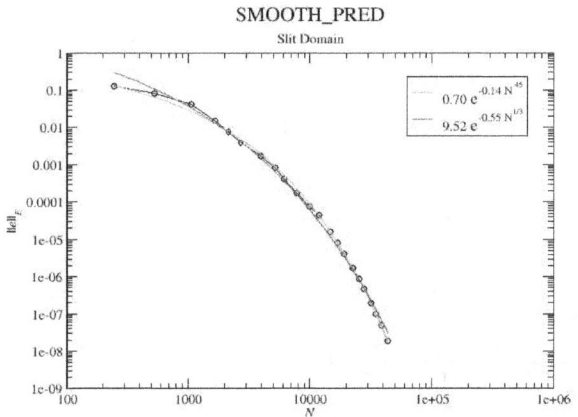

Figure 182: Log-Log plot of the convergence of the SMOOTH-PRED strategy with the slit domain problem.

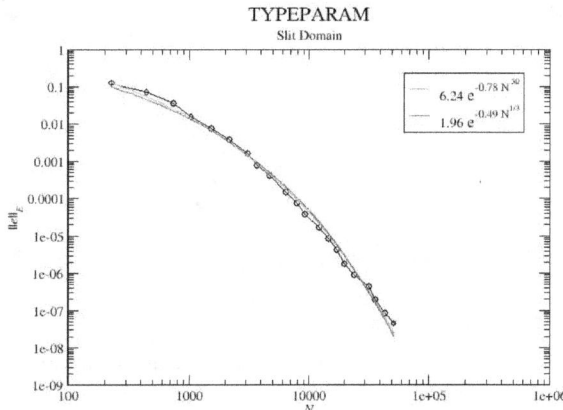

Figure 184: Log-Log plot of the convergence of the TYPEPARAM strategy with the slit domain problem.

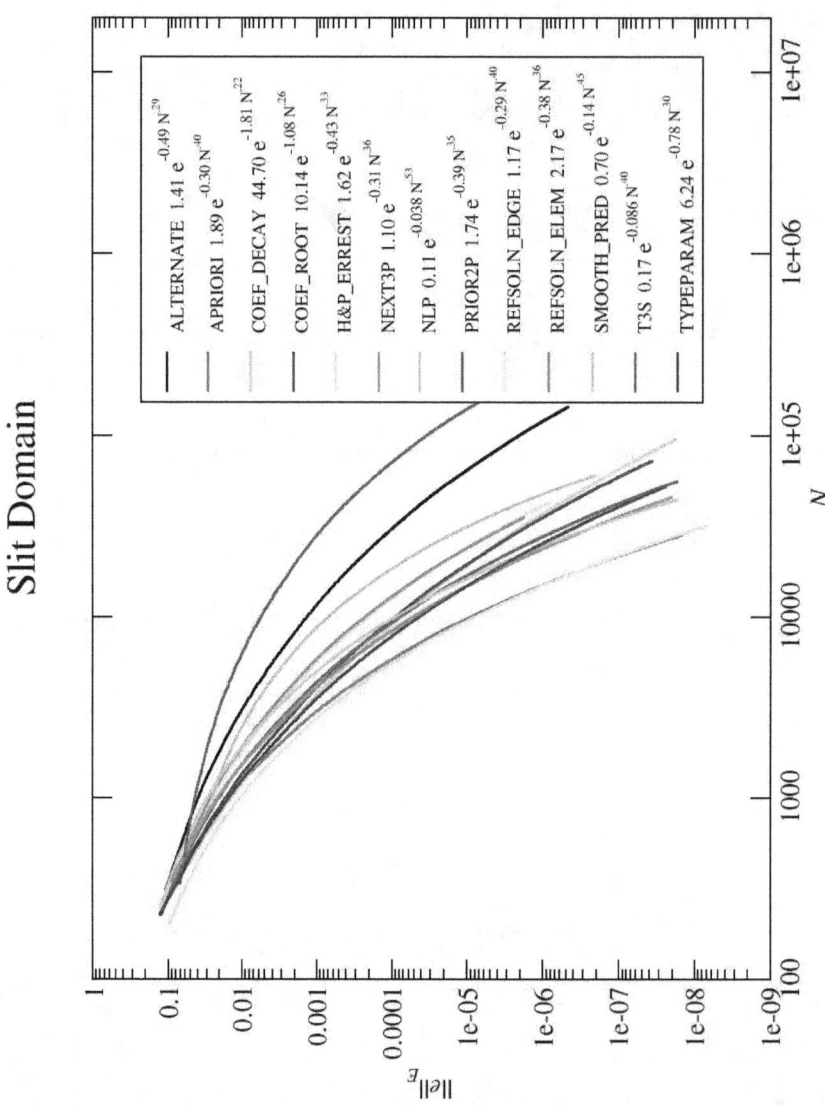

Figure 185: Log-Log plot of the convergence of all strategies with the slit domain problem.

strategy	A	B	C
ALTERNATE	1.41	-0.49	0.29
APRIORI	1.89	-0.30	0.40
COEFDECAY	44.70	-1.81	0.22
COEFROOT	10.14	-1.08	0.26
H&PERREST	1.62	-0.43	0.33
NEXT3P	1.10	-0.31	0.36
NLP	0.11	-0.038	0.53
PRIOR2P	1.74	-0.39	0.35
REFSOLNEDGE	1.17	-0.29	0.40
REFSOLNELEM	2.17	-0.38	0.36
SMOOTHPRED	0.70	-0.14	0.45
T3S	0.17	-0.086	0.40
TYPEPARAM	6.24	-0.78	0.30

Table 21: Parameters of the least squares fit for $\|e_{hp}\|_E = Ae^{BN_{dof}^C}$ for the slit domain problem.

strategy	A	B
ALTERNATE	0.54	-0.27
APRIORI	12.72	-0.67
COEFDECAY	0.57	-0.39
COEFROOT	0.97	-0.42
H&PERREST	1.54	-0.41
NEXT3P	1.74	-0.42
NLP	3.08	-0.41
PRIOR2P	2.92	-0.49
REFSOLNEDGE	7.35	-0.65
REFSOLNELEM	4.92	-0.54
SMOOTHPRED	9.52	-0.55
T3S	0.40	-0.21
TYPEPARAM	1.96	-0.49

Table 23: Parameters of the least squares fit for $\|e_{hp}\|_E = Ae^{BN_{dof}^{1/3}}$ for the slit domain problem.

strategy	factor
REFSOLNEDGE	1.00
COEFDECAY	1.18
APRIORI	1.18
TYPEPARAM	1.21
COEFROOT	1.30
REFSOLNELEM	1.48
PRIOR2P	1.48
H&PERREST	1.74
NEXT3P	1.86
SMOOTHPRED	1.86
NLP	2.41
ALTERNATE	2.99
T3S	6.11

Table 22: Factor by which N is larger than the best strategy for the slit domain problem at low accuracy, 1.0×10^{-2}.

strategy	factor
REFSOLNEDGE	1.00
APRIORI	1.01
REFSOLNELEM	1.60
SMOOTHPRED	1.65
TYPEPARAM	1.70
PRIOR2P	1.86
COEFROOT	2.37
COEFDECAY	2.48
NEXT3P	2.66
H&PERREST	2.74
NLP	3.18
ALTERNATE	8.05
T3S	15.97

Table 24: Factor by which N is larger than the best strategy for the slit domain problem at high accuracy, 1.0×10^{-6}.

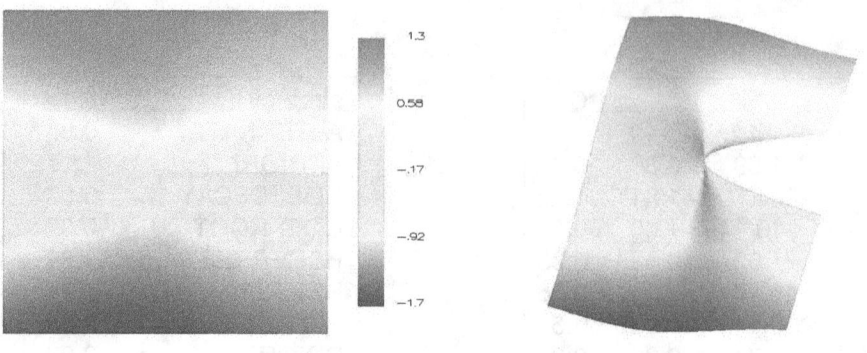

Figure 186: The u component of the solution of the mode 1 linear elasticity problem.

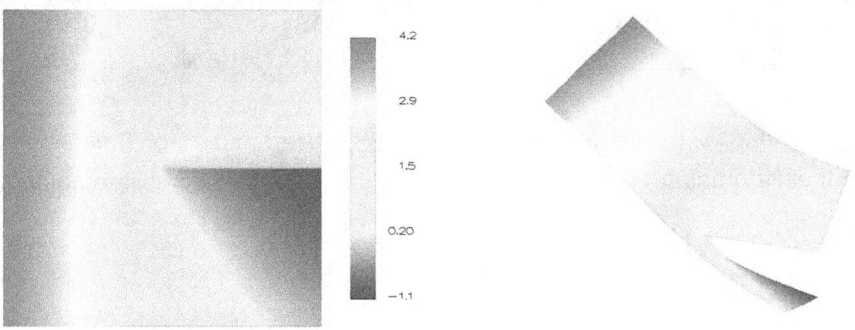

Figure 187: The v component of the solution of the mode 1 linear elasticity problem.

5.7 Linear Elasticity, Mode 1

The linear elasticity problem is a coupled system of two equations with a mixed derivative in the coupling term and different coefficients on the second order x and y terms. The domain is a square with a slit, as in the reentrant corner slit domain problem (Section 5.6). The boundary conditions are Dirichlet. For further details, see [21]. We consider two solutions, refered to as mode 1 and mode 2, by using different boundary conditions. Both solutions have a singularity at the origin, with the mode 1 solution having the stronger singularity. This section contains the results for the mode 1 solution. $\tau = 10^{-3}$ for the grid images. The APRIORI strategy refines by h if the element contains the origin and by p otherwise.

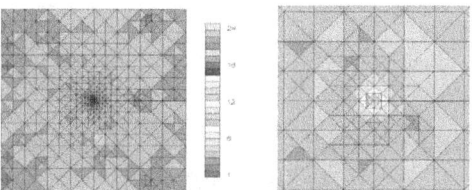

Figure 188: Example grid for the ALTERNATE strategy with the mode 1 linear elasticity problem, including details at the singularity.

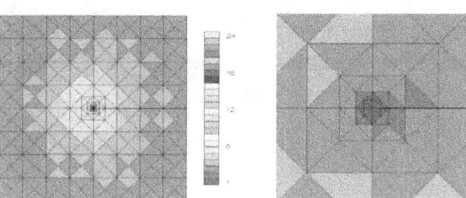

Figure 189: Example grid for the APRIORI strategy with the mode 1 linear elasticity problem, including details at the singularity.

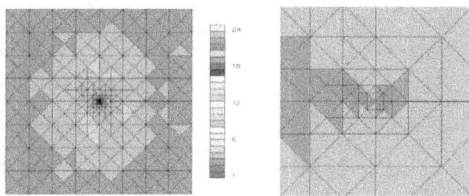

Figure 190: Example grid for the COEFDECAY strategy with the mode 1 linear elasticity problem, including details at the singularity.

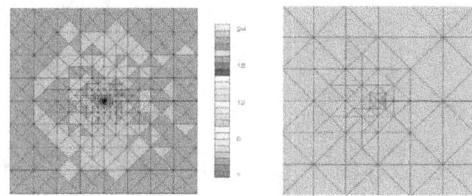

Figure 191: Example grid for the COEFROOT strategy with the mode 1 linear elasticity problem, including details at the singularity.

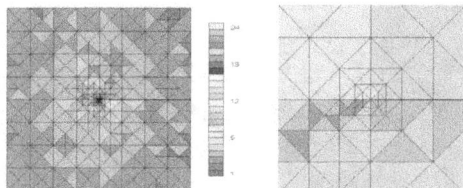

Figure 192: Example grid for the H&PERREST strategy with the mode 1 linear elasticity problem, including details at the singularity.

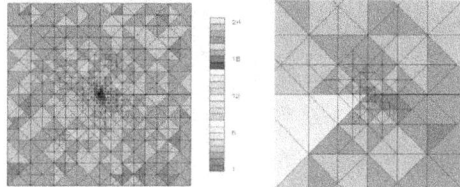

Figure 193: Example grid for the NEXT3P strategy with the mode 1 linear elasticity problem, including details at the singularity.

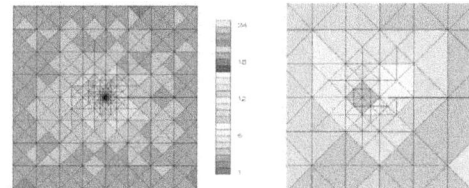

Figure 194: Example grid for the NLP strategy with the mode 1 linear elasticity problem, including details at the singularity.

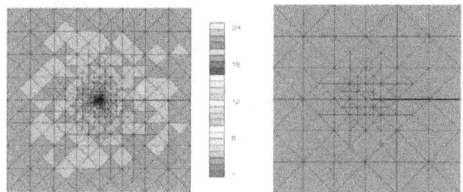

Figure 195: Example grid for the PRIOR2P strategy with the mode 1 linear elasticity problem, including details at the singularity.

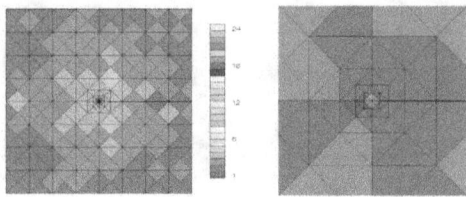

Figure 196: Example grid for the REFSOLNEDGE strategy with the mode 1 linear elasticity problem, including details at the singularity.

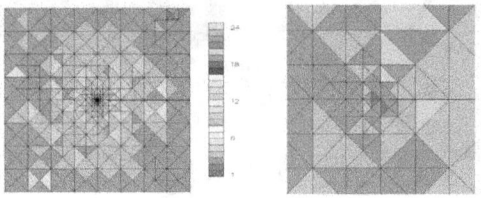

Figure 197: Example grid for the REFSOLNELEM strategy with the mode 1 linear elasticity problem, including details at the singularity.

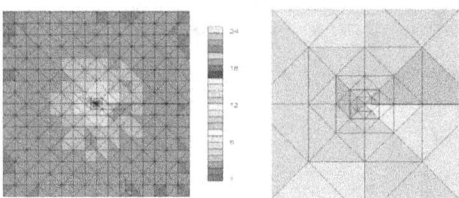

Figure 198: Example grid for the SMOOTHPRED strategy with the mode 1 linear elasticity problem, including details at the singularity.

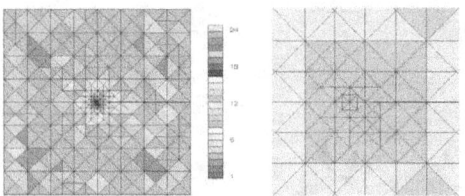

Figure 199: Example grid for the T3S strategy with the mode 1 linear elasticity problem, including details at the singularity.

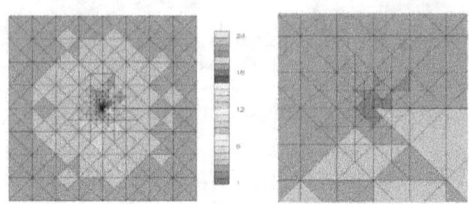

Figure 200: Example grid for the TYPEPARAM strategy with the mode 1 linear elasticity problem, including details at the singularity.

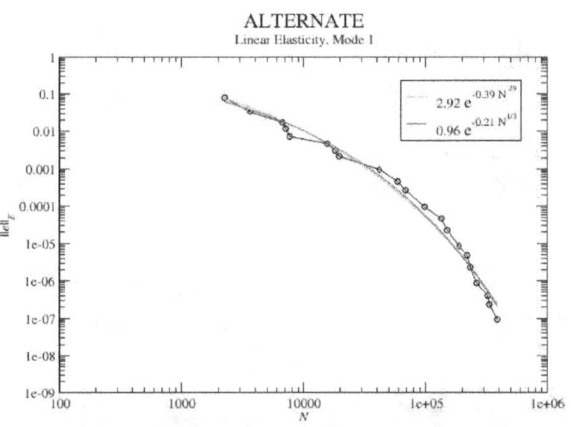

Figure 201: Log-Log plot of the convergence of the ALTERNATE strategy with the mode 1 linear elasticity problem.

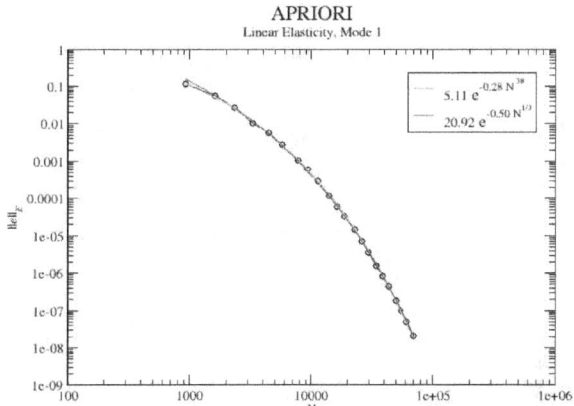

Figure 202: Log-Log plot of the convergence of the APRIORI strategy with the mode 1 linear elasticity problem.

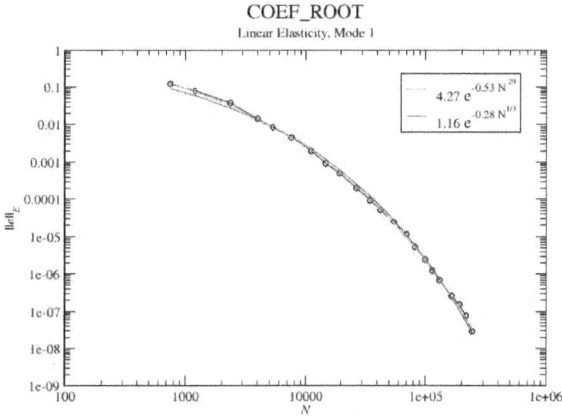

Figure 204: Log-Log plot of the convergence of the COEF-ROOT strategy with the mode 1 linear elasticity problem.

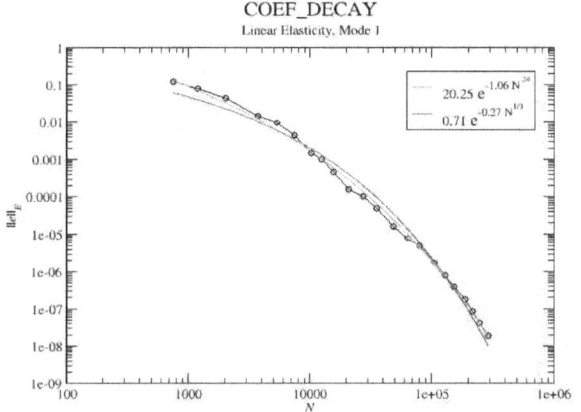

Figure 203: Log-Log plot of the convergence of the COEF-DECAY strategy with the mode 1 linear elasticity problem.

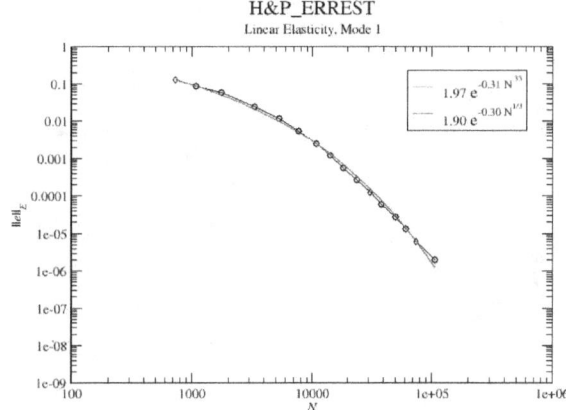

Figure 205: Log-Log plot of the convergence of the H&P-ERREST strategy with the mode 1 linear elasticity problem.

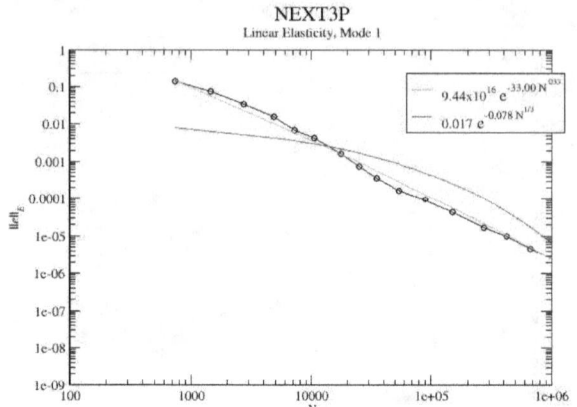

Figure 206: Log-Log plot of the convergence of the NEXT3P strategy with the mode 1 linear elasticity problem.

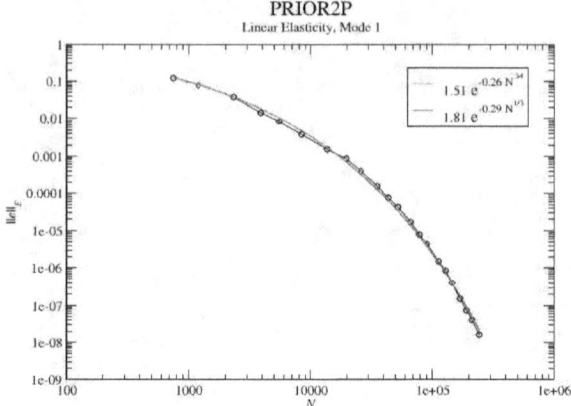

Figure 208: Log-Log plot of the convergence of the PRIOR2P strategy with the mode 1 linear elasticity problem.

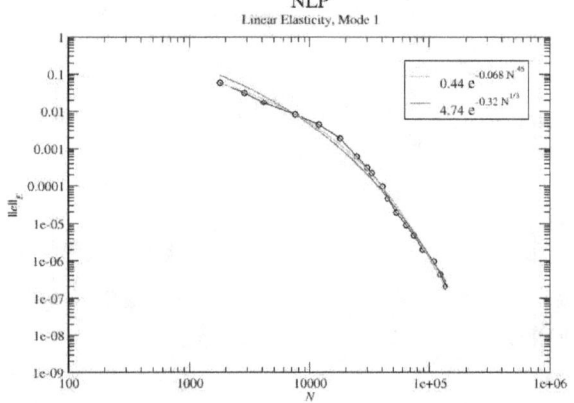

Figure 207: Log-Log plot of the convergence of the NLP strategy with the mode 1 linear elasticity problem.

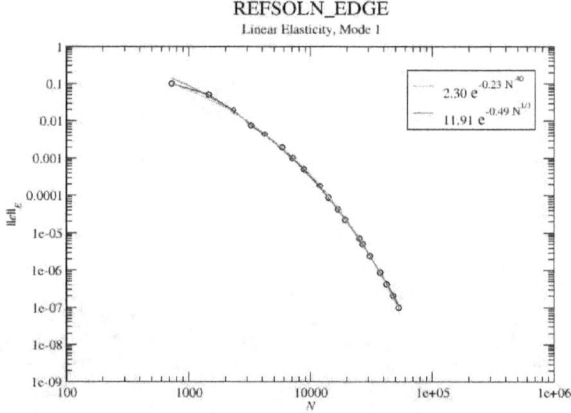

Figure 209: Log-Log plot of the convergence of the REFSOLN-EDGE strategy with the mode 1 linear elasticity problem.

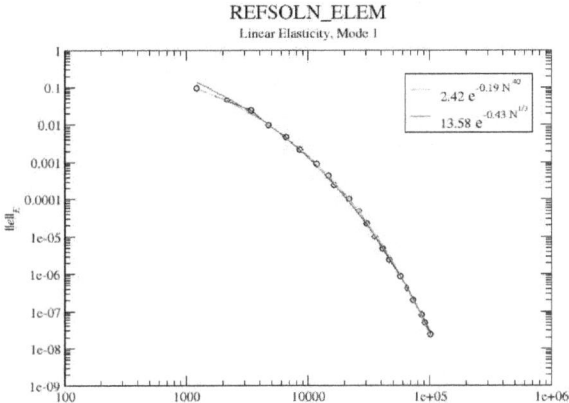

Figure 210: Log-Log plot of the convergence of the REFSOLN-ELEM strategy with the mode 1 linear elasticity problem.

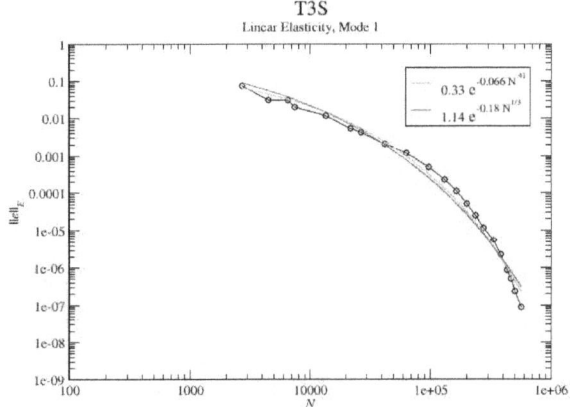

Figure 212: Log-Log plot of the convergence of the T3S strategy with the mode 1 linear elasticity problem.

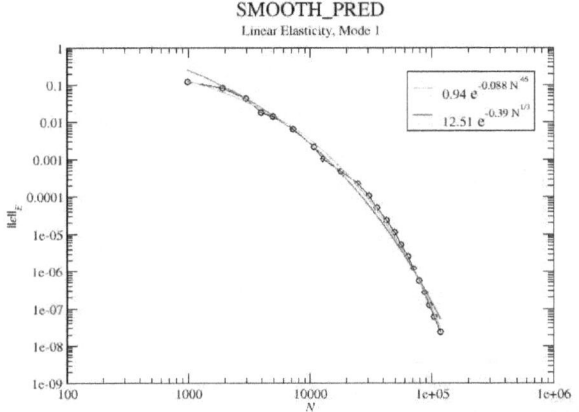

Figure 211: Log-Log plot of the convergence of the SMOOTH-PRED strategy with the mode 1 linear elasticity problem.

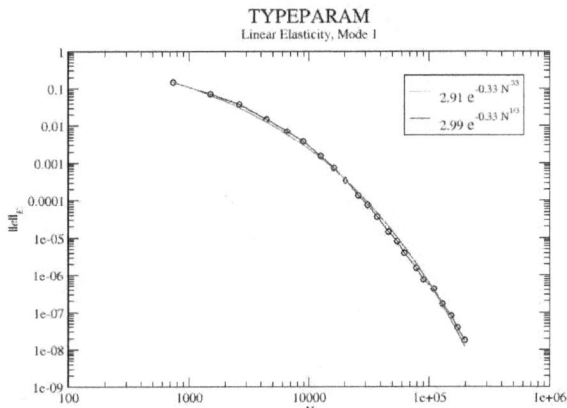

Figure 213: Log-Log plot of the convergence of the TYPEPARAM strategy with the mode 1 linear elasticity problem.

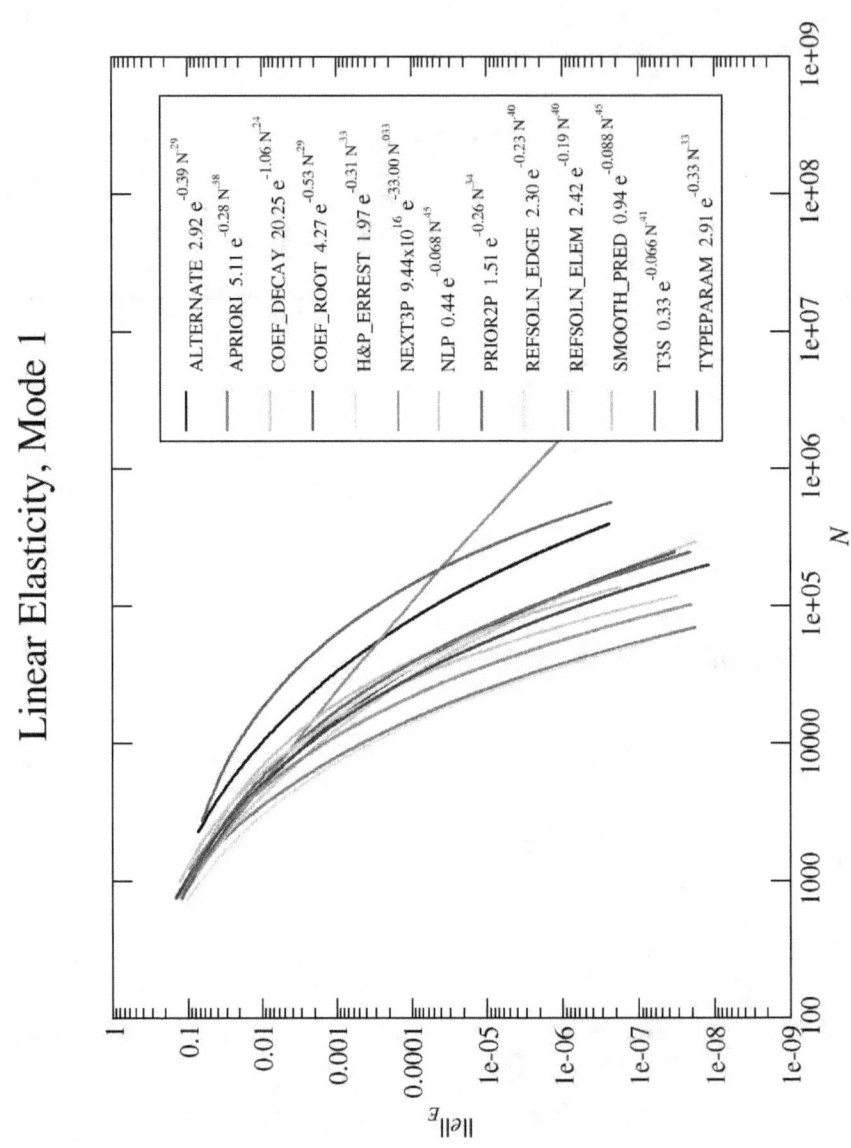

Figure 214: Log-Log plot of the convergence of all strategies with the mode 1 linear elasticity problem.

strategy	A	B	C
ALTERNATE	2.92	-0.39	0.29
APRIORI	5.11	-0.28	0.38
COEFDECAY	20.25	-1.06	0.24
COEFROOT	4.27	-0.53	0.29
H&PERREST	1.97	-0.31	0.33
NEXT3P	9.44×10^{6}	-33.00	0.033
NLP	0.44	-0.068	0.45
PRIOR2P	1.51	-0.26	0.34
REFSOLNEDGE	2.30	-0.23	0.40
REFSOLNELEM	2.42	-0.19	0.40
SMOOTHPRED	0.94	-0.088	0.45
T3S	0.33	-0.066	0.41
TYPEPARAM	2.91	-0.33	0.33

Table 25: Parameters of the least squares fit for $\|e_{hp}\|_E = Ae^{BN_{dof}^C}$ for the mode 1 linear elasticity problem.

strategy	A	B
ALTERNATE	0.96	-0.21
APRIORI	20.92	-0.50
COEFDECAY	0.71	-0.27
COEFROOT	1.16	-0.28
H&PERREST	1.90	-0.30
NEXT3P	0.017	-0.078
NLP	4.74	-0.32
PRIOR2P	1.81	-0.29
REFSOLNEDGE	11.91	-0.49
REFSOLNELEM	13.58	-0.43
SMOOTHPRED	12.51	-0.39
T3S	1.14	-0.18
TYPEPARAM	2.99	-0.33

Table 27: Parameters of the least squares fit for $\|e_{hp}\|_E = Ae^{BN_{dof}^{1/3}}$ for the mode 1 linear elasticity problem.

strategy	factor
REFSOLNEDGE	1.00
APRIORI	1.17
COEFDECAY	1.39
REFSOLNELEM	1.56
COEFROOT	1.63
NEXT3P	1.69
TYPEPARAM	1.70
H&PERREST	1.77
PRIOR2P	1.90
SMOOTHPRED	2.06
NLP	2.29
ALTERNATE	3.43
T3S	6.06

Table 26: Factor by which N is larger than the best strategy for the mode 1 linear elasticity problem at low accuracy, 1.0×10^{-2}.

strategy	factor
REFSOLNEDGE	1.00
APRIORI	1.04
REFSOLNELEM	1.55
SMOOTHPRED	1.97
TYPEPARAM	2.51
NLP	2.85
H&PERREST	3.06
COEFDECAY	3.24
PRIOR2P	3.33
COEFROOT	3.36
ALTERNATE	7.90
T3S	11.93
NEXT3P	46.52

Table 28: Factor by which N is larger than the best strategy for the mode 1 linear elasticity problem at high accuracy, 1.0×10^{-6}.

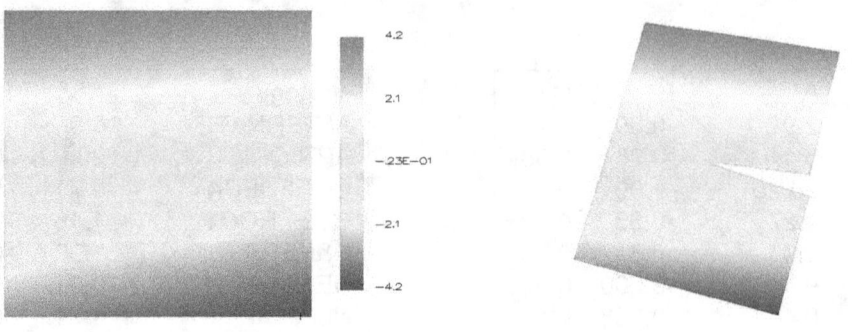

Figure 215: The u component of the solution of the mode 2 linear elasticity problem.

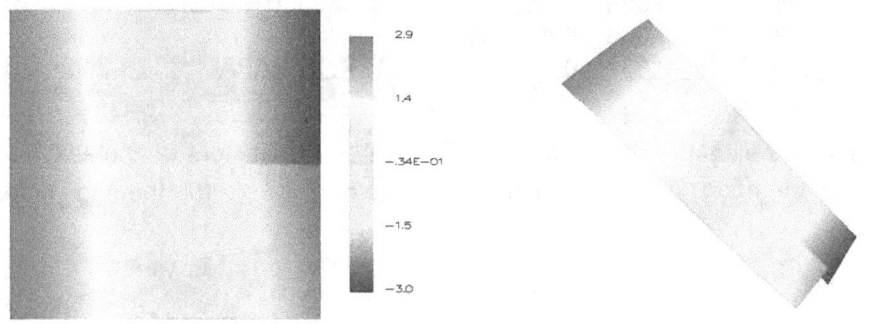

Figure 216: The v component of the solution of the mode 2 linear elasticity problem.

5.8 Linear Elasticity, Mode 2

This is the mode 2 solution of the linear elasticity problem (Section 5.7). $\tau=10^{-3}$ for the grid images. The APRIORI strategy refines by h if the element contains the origin and by p otherwise.

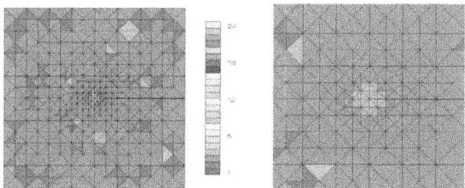

Figure 217: Example grid for the ALTERNATE strategy with the mode 2 linear elasticity problem, including details at the singularity.

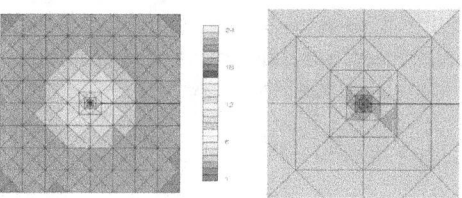

Figure 218: Example grid for the APRIORI strategy with the mode 2 linear elasticity problem, including details at the singularity.

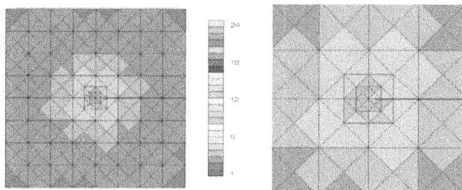

Figure 219: Example grid for the COEFDECAY strategy with the mode 2 linear elasticity problem, including details at the singularity.

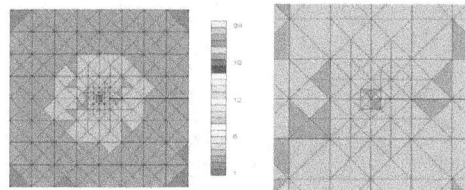

Figure 220: Example grid for the COEFROOT strategy with the mode 2 linear elasticity problem, including details at the singularity.

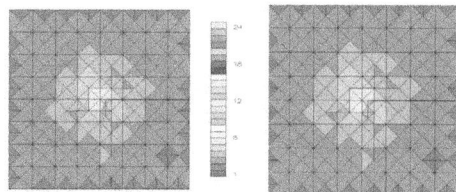

Figure 221: Example grid for the H&PERREST strategy with the mode 2 linear elasticity problem, including details at the singularity.

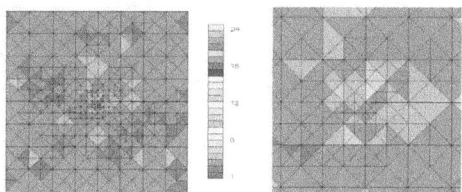

Figure 222: Example grid for the NEXT3P strategy with the mode 2 linear elasticity problem, including details at the singularity.

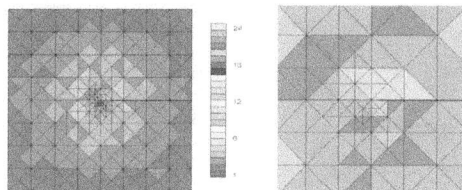

Figure 223: Example grid for the NLP strategy with the mode 2 linear elasticity problem, including details at the singularity.

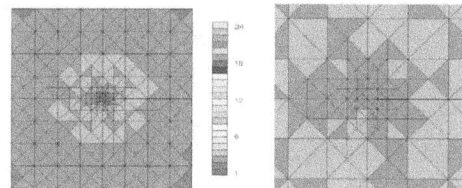

Figure 224: Example grid for the PRIOR2P strategy with the mode 2 linear elasticity problem, including details at the singularity.

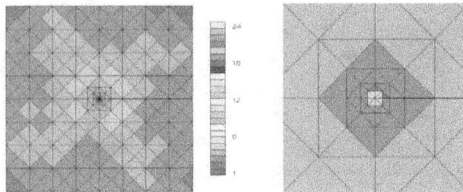

Figure 225: Example grid for the REFSOLNEDGE strategy with the mode 2 linear elasticity problem, including details at the singularity.

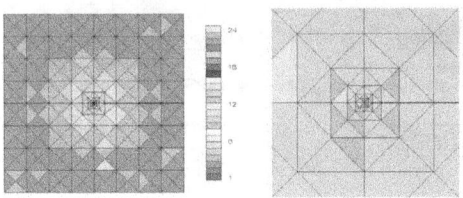

Figure 226: Example grid for the REFSOLNELEM strategy with the mode 2 linear elasticity problem, including details at the singularity.

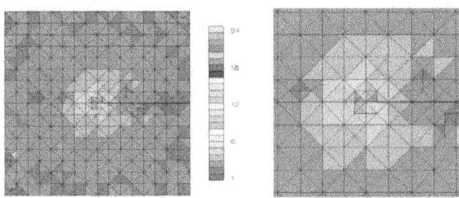

Figure 227: Example grid for the SMOOTHPRED strategy with the mode 2 linear elasticity problem, including details at the singularity.

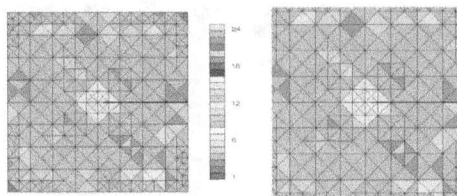

Figure 228: Example grid for the T3S strategy with the mode 2 linear elasticity problem, including details at the singularity.

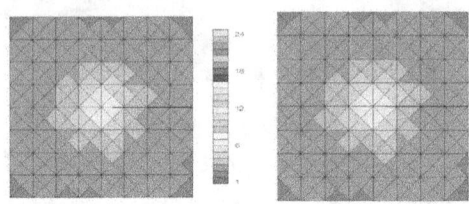

Figure 229: Example grid for the TYPEPARAM strategy with the mode 2 linear elasticity problem, including details at the singularity.

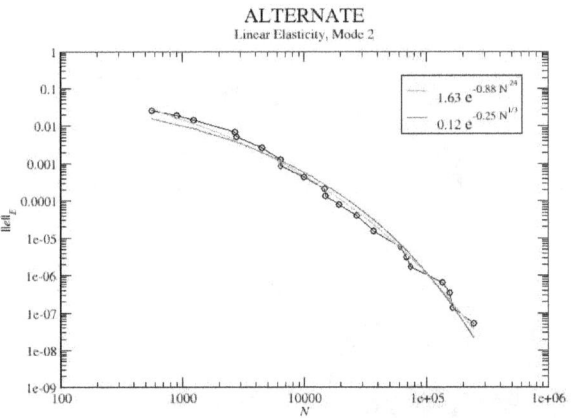

Figure 230: Log-Log plot of the convergence of the ALTERNATE strategy with the mode 2 linear elasticity problem.

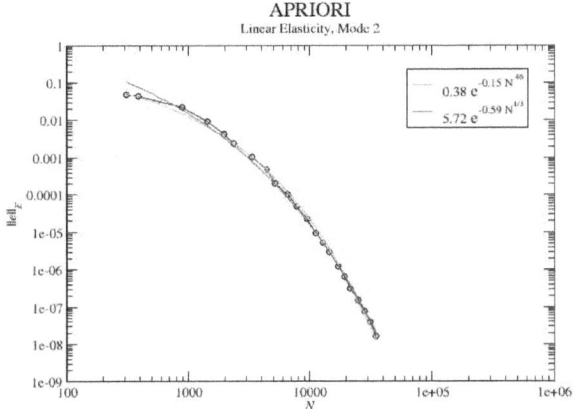

Figure 231: Log-Log plot of the convergence of the APRIORI strategy with the mode 2 linear elasticity problem.

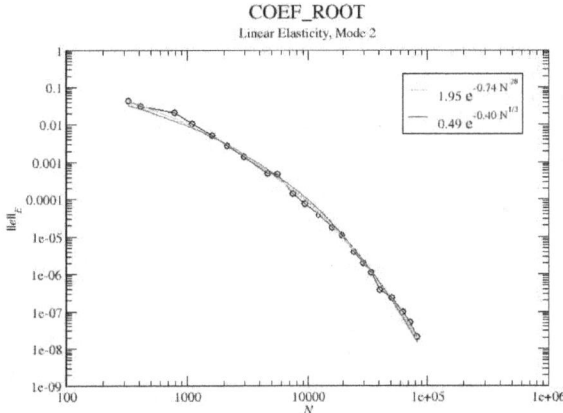

Figure 233: Log-Log plot of the convergence of the COEF-ROOT strategy with the mode 2 linear elasticity problem.

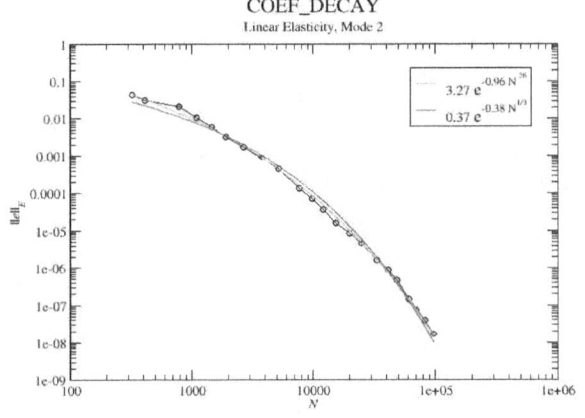

Figure 232: Log-Log plot of the convergence of the COEF-DECAY strategy with the mode 2 linear elasticity problem.

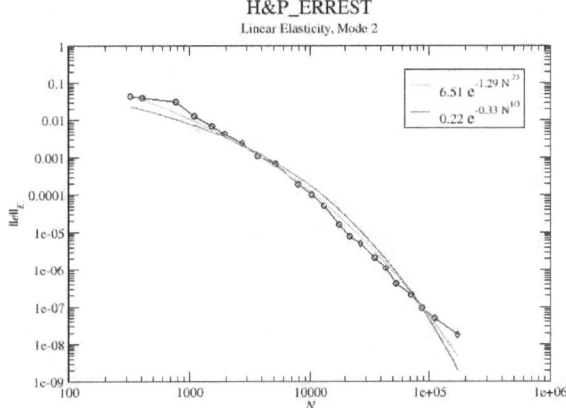

Figure 234: Log-Log plot of the convergence of the H&P-ERREST strategy with the mode 2 linear elasticity problem.

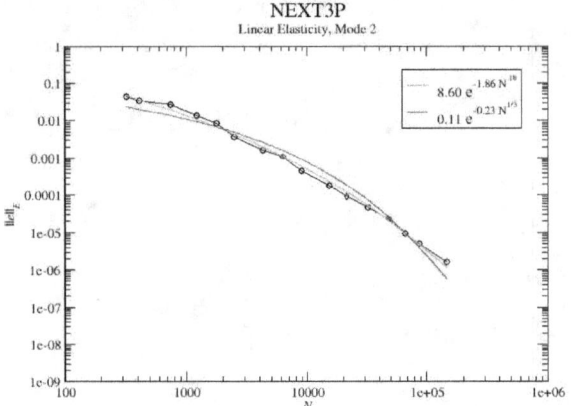

Figure 235: Log-Log plot of the convergence of the NEXT3P strategy with the mode 2 linear elasticity problem.

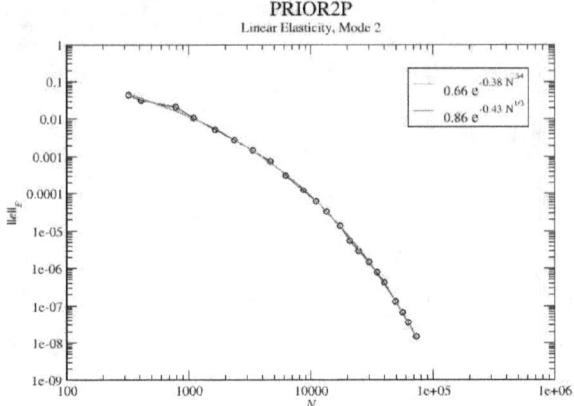

Figure 237: Log-Log plot of the convergence of the PRIOR2P strategy with the mode 2 linear elasticity problem.

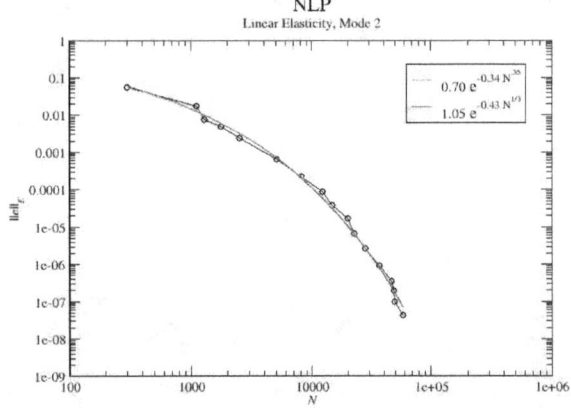

Figure 236: Log-Log plot of the convergence of the NLP strategy with the mode 2 linear elasticity problem.

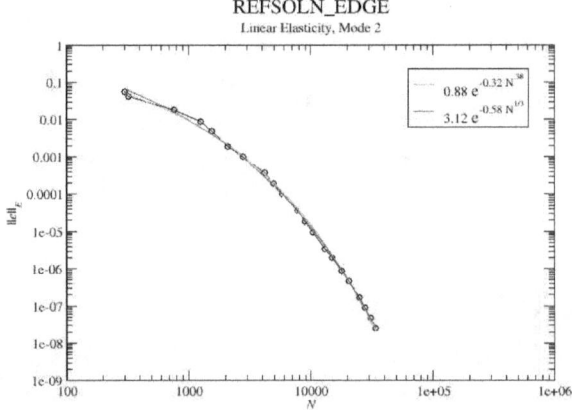

Figure 238: Log-Log plot of the convergence of the REFSOLN-EDGE strategy with the mode 2 linear elasticity problem.

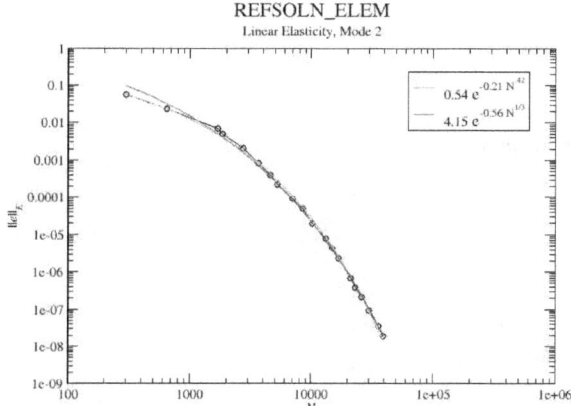
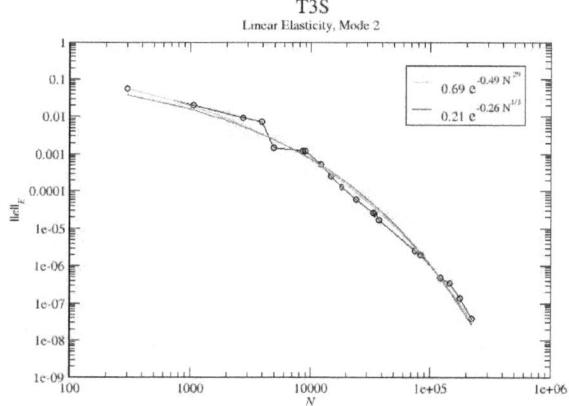

Figure 239: Log-Log plot of the convergence of the REFSOLN-ELEM strategy with the mode 2 linear elasticity problem.

Figure 241: Log-Log plot of the convergence of the T3S strategy with the mode 2 linear elasticity problem.

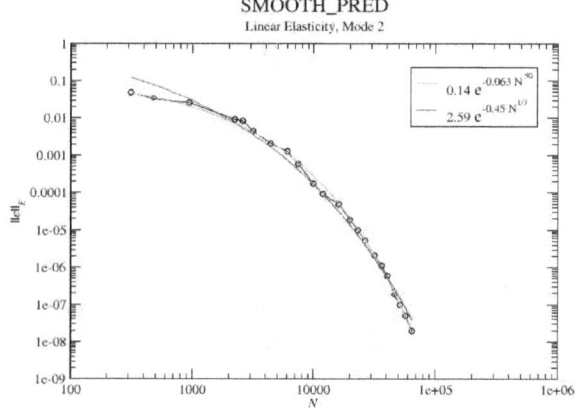
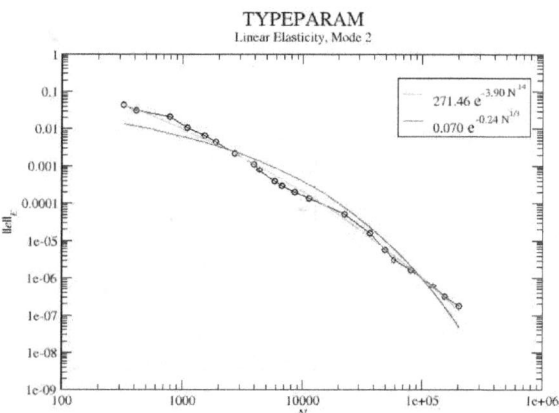

Figure 240: Log-Log plot of the convergence of the SMOOTH-PRED strategy with the mode 2 linear elasticity problem.

Figure 242: Log-Log plot of the convergence of the TYPEPARAM strategy with the mode 2 linear elasticity problem.

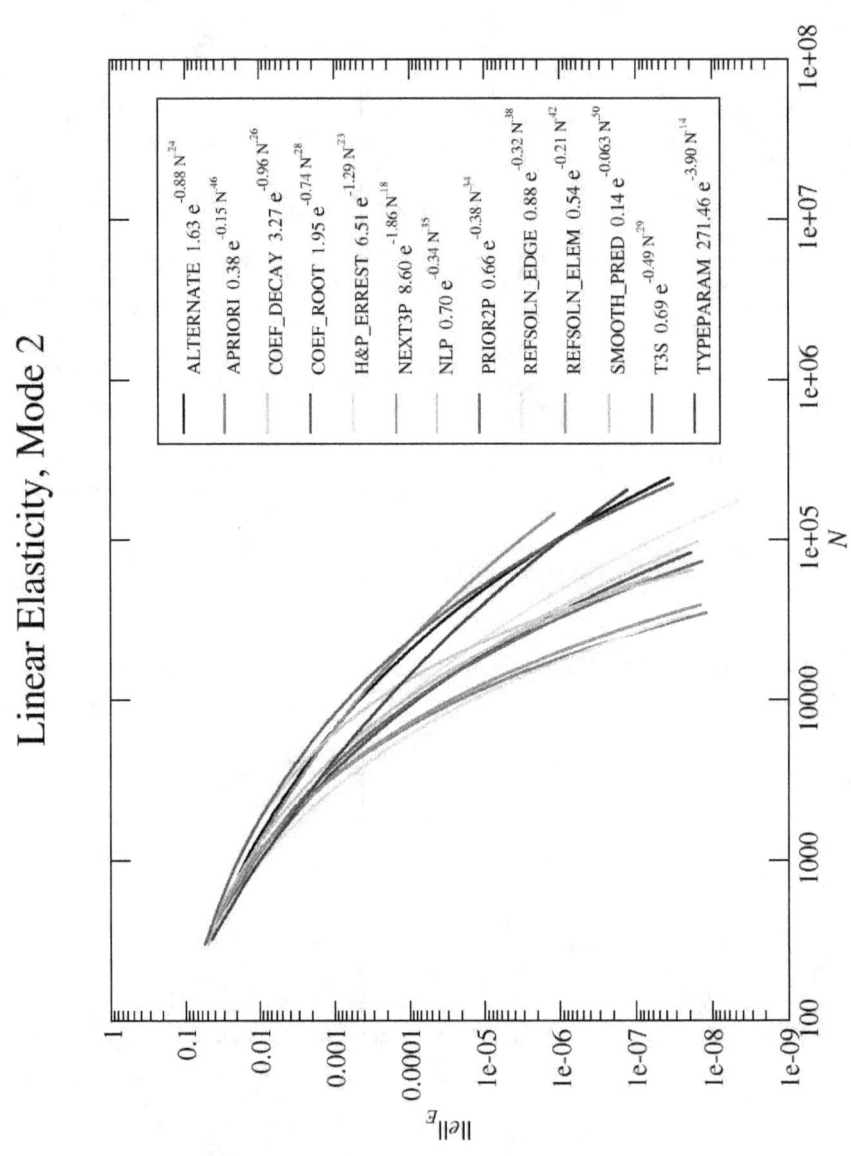

Figure 243: Log-Log plot of the convergence of all strategies with the mode 2 linear elasticity problem.

strategy	A	B	C
ALTERNATE	1.63	-0.88	0.24
APRIORI	0.38	-0.15	0.46
COEFDECAY	3.27	-0.96	0.26
COEFROOT	1.95	-0.74	0.28
H&PERREST	6.51	-1.29	0.23
NEXT3P	8.60	-1.86	0.18
NLP	0.70	-0.34	0.35
PRIOR2P	0.66	-0.38	0.34
REFSOLNEDGE	0.88	-0.32	0.38
REFSOLNELEM	0.54	-0.21	0.42
SMOOTHPRED	0.14	-0.063	0.50
T3S	0.69	-0.49	0.29
TYPEPARAM	271.46	-3.90	0.14

Table 29: Parameters of the least squares fit for $\|e_{hp}\|_E = Ae^{BN_{dof}^C}$ for the mode 2 linear elasticity problem.

strategy	A	B
ALTERNATE	0.12	-0.25
APRIORI	5.72	-0.59
COEFDECAY	0.37	-0.38
COEFROOT	0.49	-0.40
H&PERREST	0.22	-0.33
NEXT3P	0.11	-0.23
NLP	1.05	-0.43
PRIOR2P	0.86	-0.43
REFSOLNEDGE	3.12	-0.58
REFSOLNELEM	4.15	-0.56
SMOOTHPRED	2.59	-0.45
T3S	0.21	-0.26
TYPEPARAM	0.070	-0.24

Table 31: Parameters of the least squares fit for $\|e_{hp}\|_E = Ae^{BN_{dof}^{1/3}}$ for the mode 2 linear elasticity problem.

strategy	factor
REFSOLNEDGE	1.00
TYPEPARAM	1.03
COEFDECAY	1.03
COEFROOT	1.06
H&PERREST	1.10
PRIOR2P	1.16
REFSOLNELEM	1.18
APRIORI	1.19
NLP	1.34
NEXT3P	1.35
ALTERNATE	1.48
SMOOTHPRED	1.87
T3S	1.88

Table 30: Factor by which N is larger than the best strategy for the mode 2 linear elasticity problem at low accuracy, 1.0×10^{-2}.

strategy	factor
REFSOLNEDGE	1.00
APRIORI	1.03
REFSOLNELEM	1.14
PRIOR2P	1.84
NLP	1.96
COEFROOT	2.01
SMOOTHPRED	2.04
COEFDECAY	2.15
H&PERREST	2.75
TYPEPARAM	5.66
T3S	5.74
ALTERNATE	5.81
NEXT3P	8.76

Table 32: Factor by which N is larger than the best strategy for the mode 2 linear elasticity problem at high accuracy, 1.0×10^{-6}.

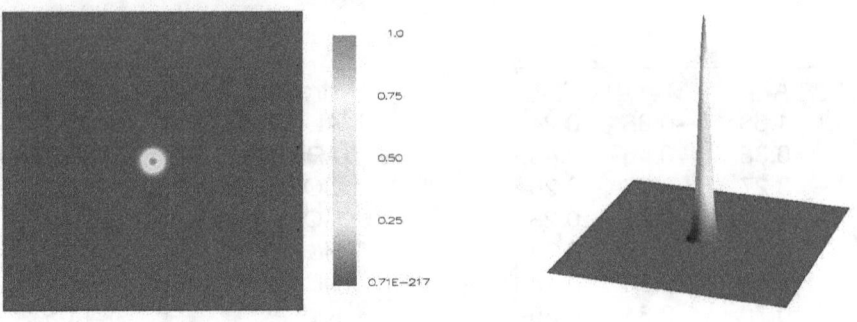

Figure 244: The solution of the mild peak problem.

5.9 Mild Peak

The peak problem contains a Gaussian peak in the interior of the domain. It is Poisson's equation on the unit square with Dirichlet boundary conditions. The solution is

$$e^{-\alpha((x-x_c)^2 + (y-y_c)^2)}$$

where (x_c, y_c) is the location of the peak, and α determines the strength of the peak. For the easy form of this problem, we use $\alpha = 1000$ and $(x_c, y_c) = (0.5, 0.5)$. For this problem, we used $\tau = 10^{-5}$ for the grid images. The APRIORI strategy refines by h if the element touches the center of the peak and by p otherwise.

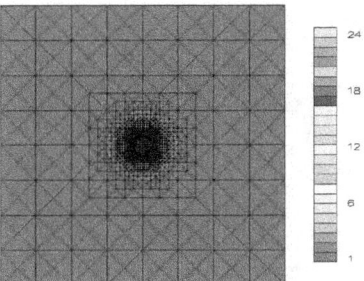

Figure 245: Example grid for the ALTERNATE strategy with the mild peak problem.

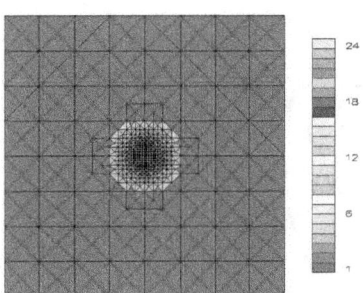

Figure 248: Example grid for the COEFROOT strategy with the mild peak problem.

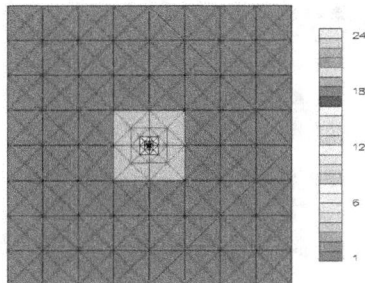

Figure 246: Example grid for the APRIORI strategy with the mild peak problem.

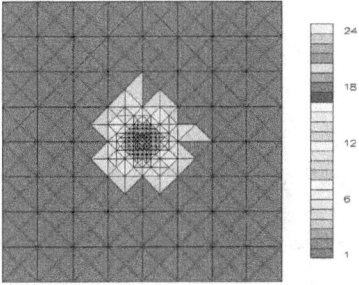

Figure 249: Example grid for the H&PERREST strategy with the mild peak problem.

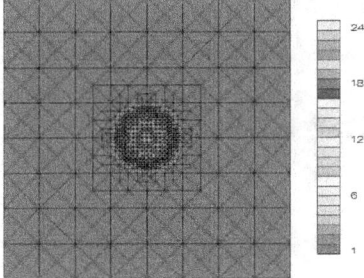

Figure 247: Example grid for the COEFDECAY strategy with the mild peak problem.

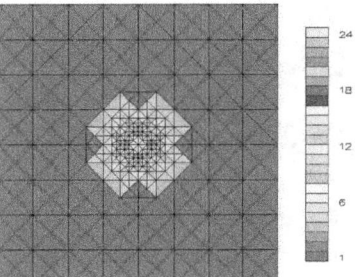

Figure 250: Example grid for the NEXT3P strategy with the mild peak problem.

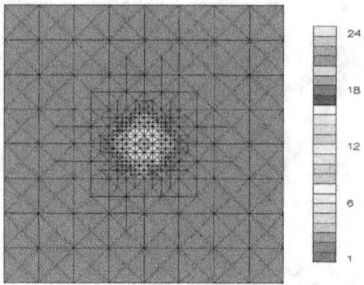

Figure 251: Example grid for the NLP strategy with the mild peak problem.

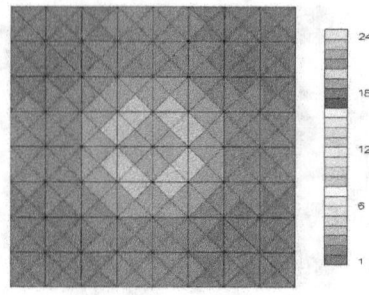

Figure 254: Example grid for the REFSOLNELEM strategy with the mild peak problem.

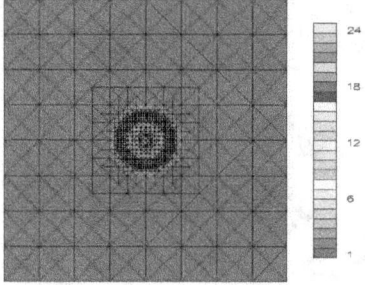

Figure 252: Example grid for the PRIOR2P strategy with the mild peak problem.

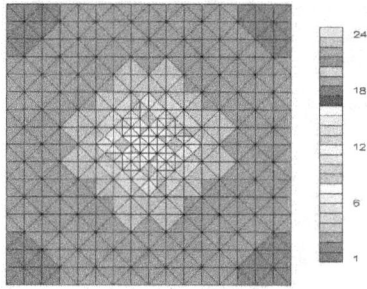

Figure 255: Example grid for the SMOOTHPRED strategy with the mild peak problem.

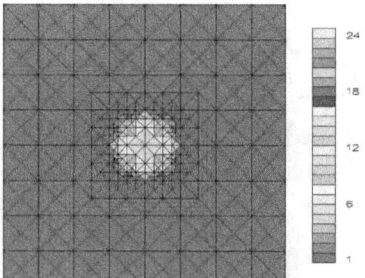

Figure 253: Example grid for the REFSOLNEDGE strategy with the mild peak problem.

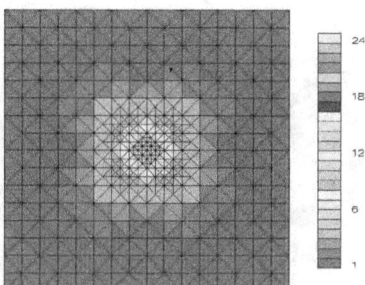

Figure 256: Example grid for the T3S strategy with the mild peak problem.

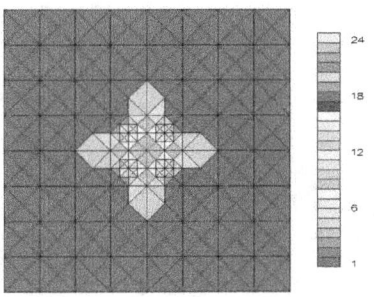

Figure 257: Example grid for the TYPEPARAM strategy with the mild peak problem.

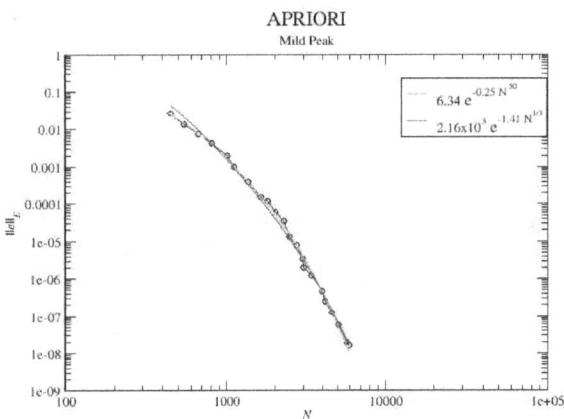

Figure 259: Log-Log plot of the convergence of the APRIORI strategy with the mild peak problem.

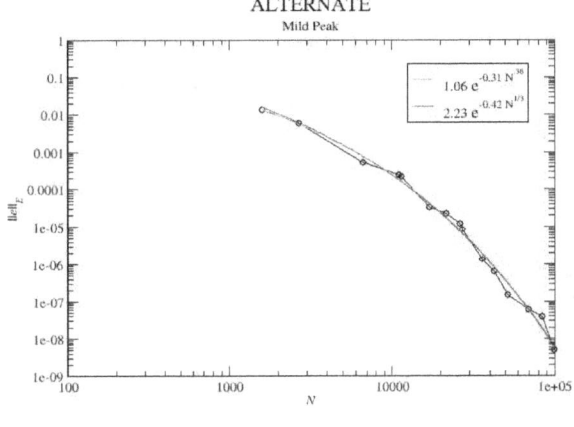

Figure 258: Log-Log plot of the convergence of the ALTERNATE strategy with the mild peak problem.

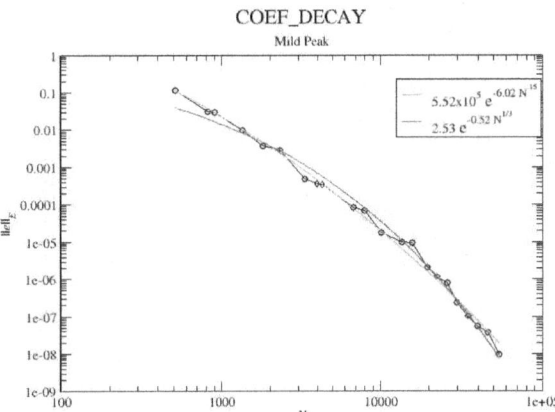

Figure 260: Log-Log plot of the convergence of the COEF_DECAY strategy with the mild peak problem.

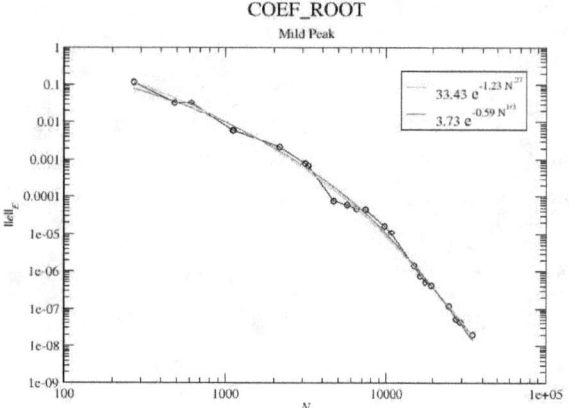
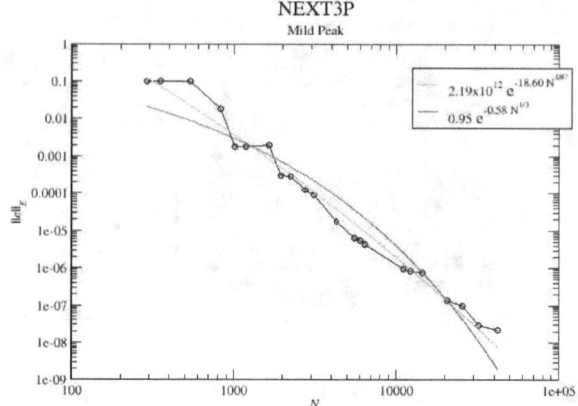

Figure 261: Log-Log plot of the convergence of the COEF-ROOT strategy with the mild peak problem.

Figure 263: Log-Log plot of the convergence of the NEXT3P strategy with the mild peak problem.

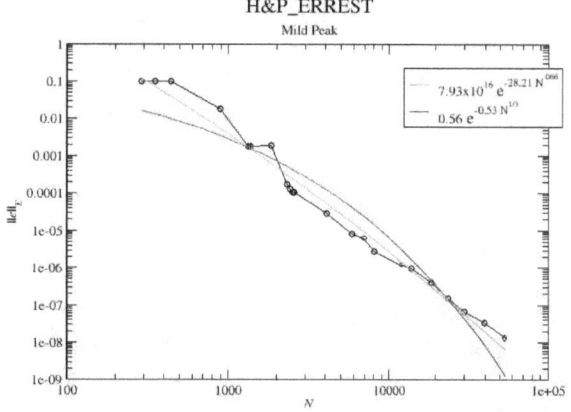
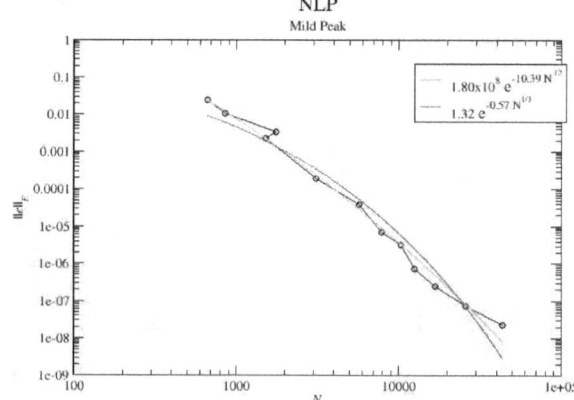

Figure 262: Log-Log plot of the convergence of the H&P-ERREST strategy with the mild peak problem.

Figure 264: Log-Log plot of the convergence of the NLP strategy with the mild peak problem.

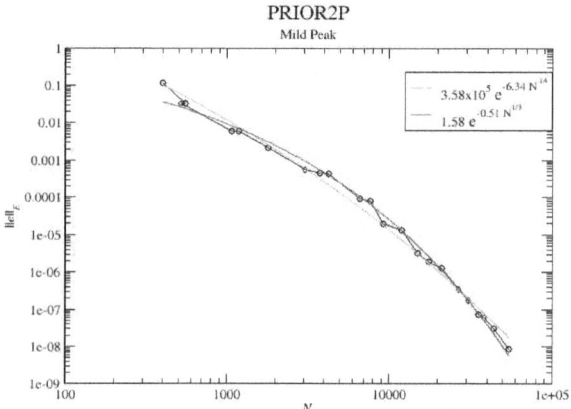

Figure 265: Log-Log plot of the convergence of the PRIOR2P strategy with the mild peak problem.

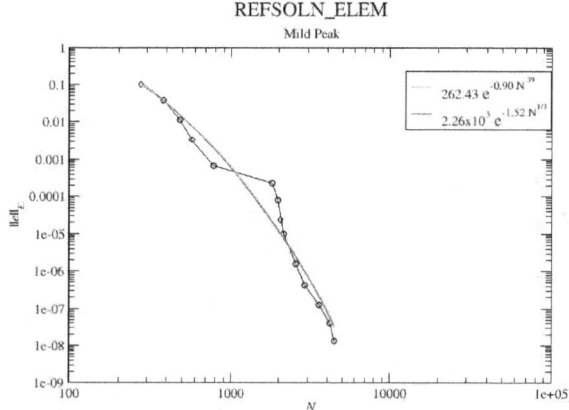

Figure 267: Log-Log plot of the convergence of the REFSOLN-ELEM strategy with the mild peak problem.

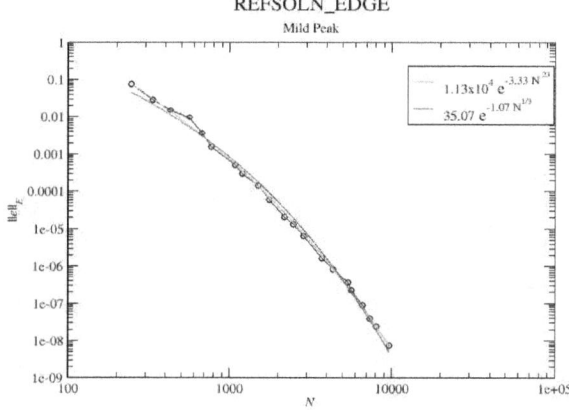

Figure 266: Log-Log plot of the convergence of the REFSOLN-EDGE strategy with the mild peak problem.

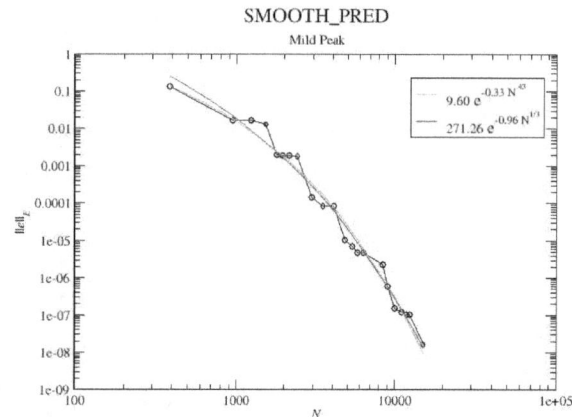

Figure 268: Log-Log plot of the convergence of the SMOOTH-PRED strategy with the mild peak problem.

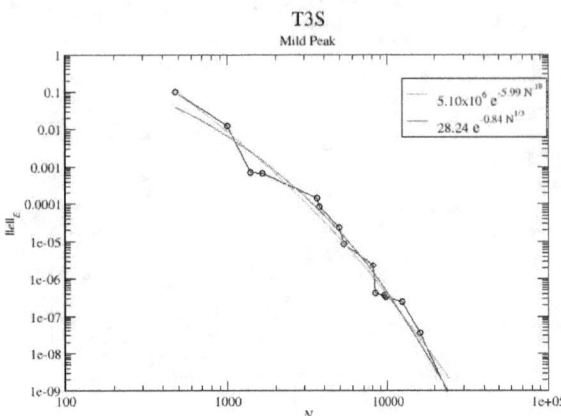

Figure 269: Log-Log plot of the convergence of the T3S strategy with the mild peak problem.

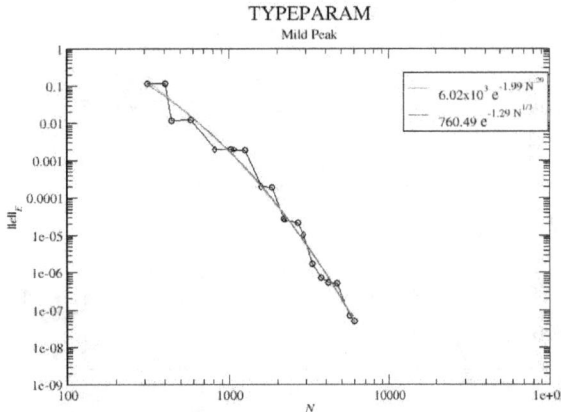

Figure 270: Log-Log plot of the convergence of the TYPEPARAM strategy with the mild peak problem.

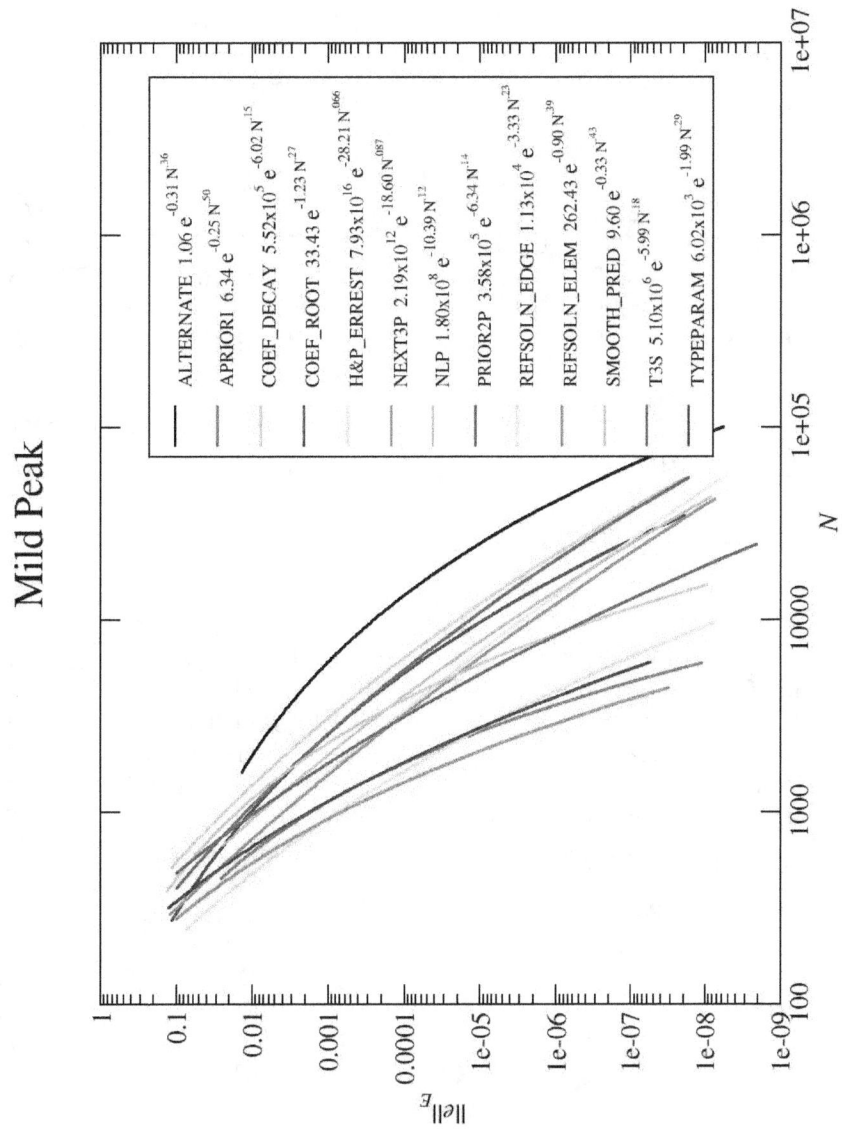

Figure 271: Log-Log plot of the convergence of all strategies with the mild peak problem.

strategy	A	B	C
ALTERNATE	1.06	-0.31	0.36
APRIORI	6.34	-0.25	0.50
COEFDECAY	5.52×10^5	-6.02	0.15
COEFROOT	33.43	-1.23	0.27
H&PERREST	7.93×10^{6}	-28.21	0.066
NEXT3P	2.19×10^{12}	-18.60	0.087
NLP	1.80×10^{8}	-10.39	0.12
PRIOR2P	3.58×10^{5}	-6.34	0.14
REFSOLNEDGE	1.13×10^{4}	-3.33	0.23
REFSOLNELEM	262.43	-0.90	0.39
SMOOTHPRED	9.60	-0.33	0.43
T3S	5.10×10^{6}	-5.99	0.18
TYPEPARAM	6.02×10^{3}	-1.99	0.29

Table 33: Parameters of the least squares fit for $\|e_{hp}\|_E = Ae^{BN} dof^C$ for the mild peak problem.

strategy	A	B
ALTERNATE	2.23	-0.42
APRIORI	2.16×10^{3}	-1.41
COEFDECAY	2.53	-0.52
COEFROOT	3.73	-0.59
H&PERREST	0.56	-0.53
NEXT3P	0.95	-0.58
NLP	1.32	-0.57
PRIOR2P	1.58	-0.51
REFSOLNEDGE	35.07	-1.07
REFSOLNELEM	2.26×10^{3}	-1.52
SMOOTHPRED	271.26	-0.96
T3S	28.24	-0.84
TYPEPARAM	760.49	-1.29

Table 35: Parameters of the least squares fit for $\|e_{hp}\|_E = Ae^{BN} dof^{1/3}$ for the mild peak problem.

strategy	factor
REFSOLNEDGE	1.00
REFSOLNELEM	1.13
APRIORI	1.29
TYPEPARAM	1.38
NEXT3P	1.51
H&PERREST	1.52
NLP	1.92
T3S	2.02
COEFROOT	2.13
PRIOR2P	2.26
SMOOTHPRED	2.45
COEFDECAY	2.88
ALTERNATE	4.11

Table 34: Factor by which N is larger than the best strategy for the mild peak problem at low accuracy, 1.0×10^{-2}.

strategy	factor
REFSOLNELEM	1.00
APRIORI	1.25
TYPEPARAM	1.37
REFSOLNEDGE	1.47
T3S	2.88
SMOOTHPRED	2.91
NEXT3P	4.19
H&PERREST	4.66
NLP	4.88
COEFROOT	5.67
PRIOR2P	7.09
COEFDECAY	7.60
ALTERNATE	14.36

Table 36: Factor by which N is larger than the best strategy for the mild peak problem at high accuracy, 1.0×10^{-6}.

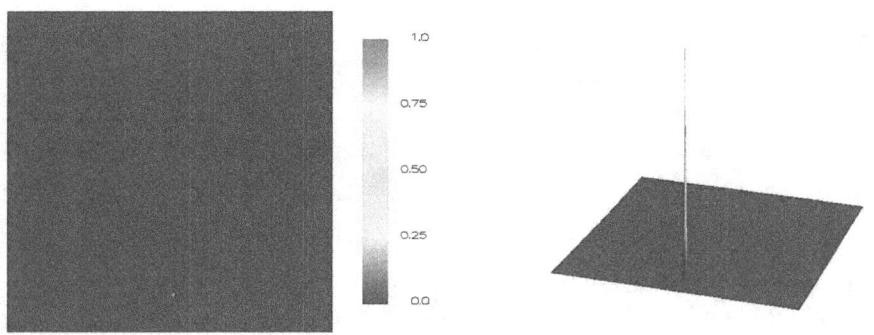

Figure 272: The solution of the sharp peak problem.

5.10 SharpPeak

This is the hard version of the peak problem (Section 5.9) with $\alpha = 100000$ and $(x_c, y_c) = (.51, .117)$. We used $\tau = 10^{-5}$ for the grid images. The APRIORI strategy refines by h if the element touches the center of the peak and by p otherwise.

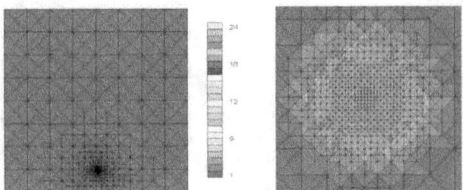

Figure 273: Example grid for the ALTERNATE strategy with the sharp peak problem, including details at the peak.

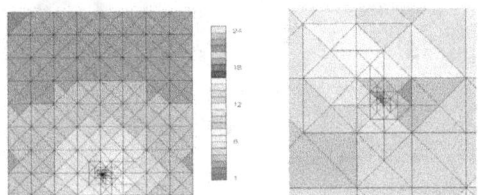

Figure 274: Example grid for the APRIORI strategy with the sharp peak problem, including details at the peak.

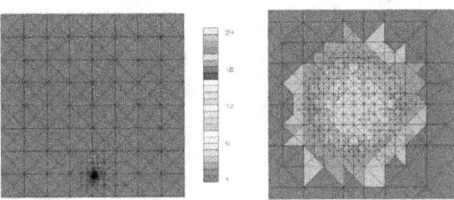

Figure 275: Example grid for the COEFDECAY strategy with the sharp peak problem, including details at the peak.

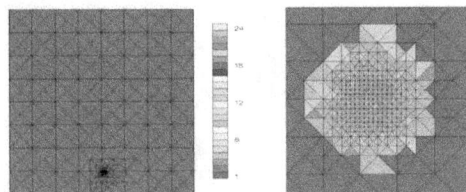

Figure 276: Example grid for the COEFROOT strategy with the sharp peak problem, including details at the peak.

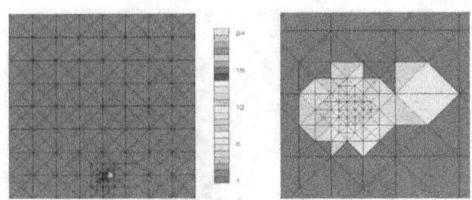

Figure 277: Example grid for the H&PERREST strategy with the sharp peak problem, including details at the peak.

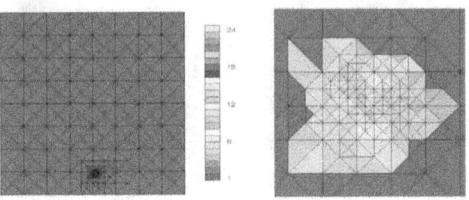

Figure 278: Example grid for the NEXT3P strategy with the sharp peak problem, including details at the peak.

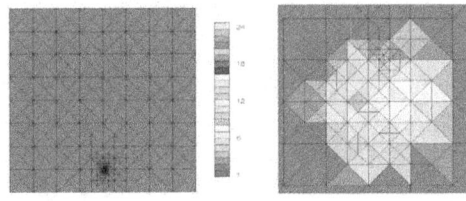

Figure 279: Example grid for the NLP strategy with the sharp peak problem, including details at the peak.

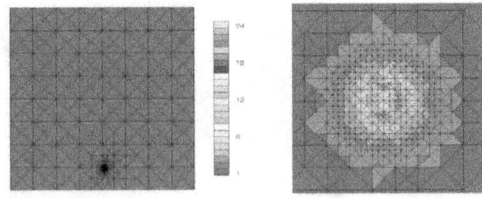

Figure 280: Example grid for the PRIOR2P strategy with the sharp peak problem, including details at the peak.

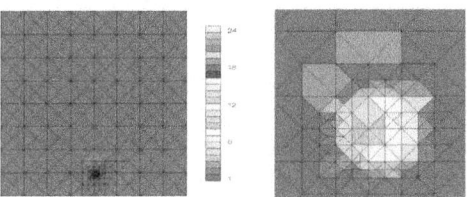

Figure 281: Example grid for the REFSOLNEDGE strategy with the sharp peak problem, including details at the peak.

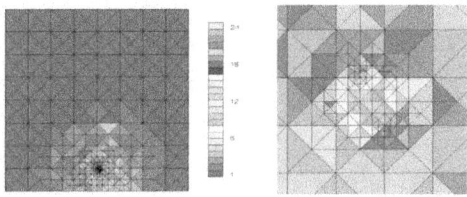

Figure 282: Example grid for the REFSOLNELEM strategy with the sharp peak problem, including details at the peak.

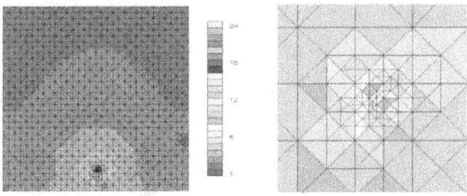

Figure 283: Example grid for the SMOOTHPRED strategy with the sharp peak problem, including details at the peak.

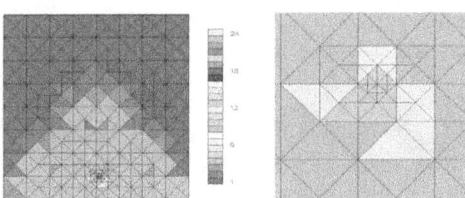

Figure 284: Example grid for the T3S strategy with the sharp peak problem, including details at the peak.

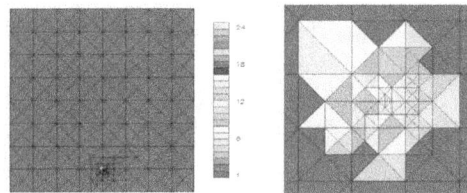

Figure 285: Example grid for the TYPEPARAM strategy with the sharp peak problem, including details at the peak.

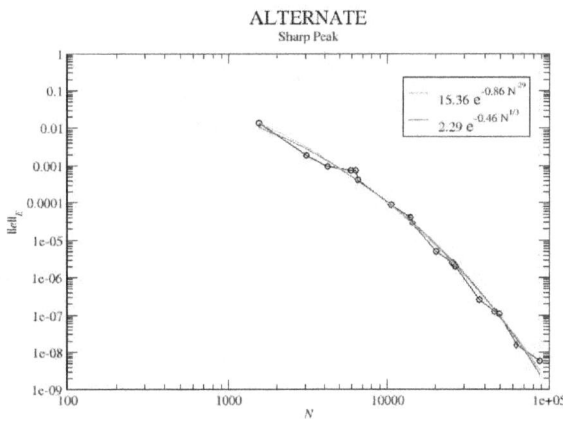

Figure 286: Log-Log plot of the convergence of the ALTERNATE strategy with the sharp peak problem.

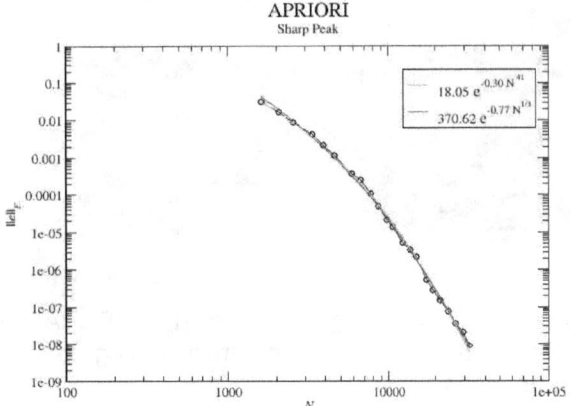
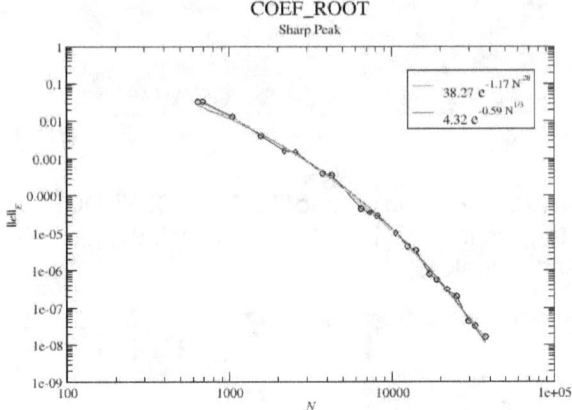

Figure 287: Log-Log plot of the convergence of the APRIORI strategy with the sharp peak problem.

Figure 289: Log-Log plot of the convergence of the COEF-ROOT strategy with the sharp peak problem.

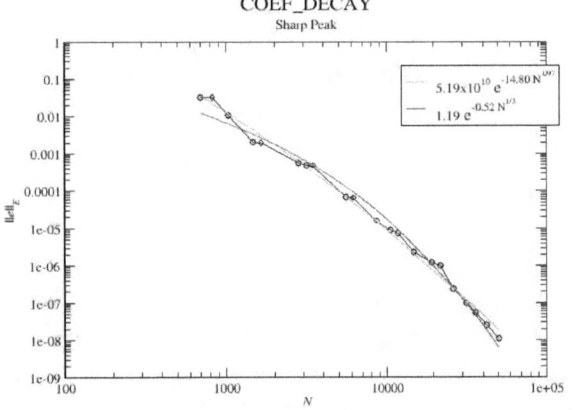
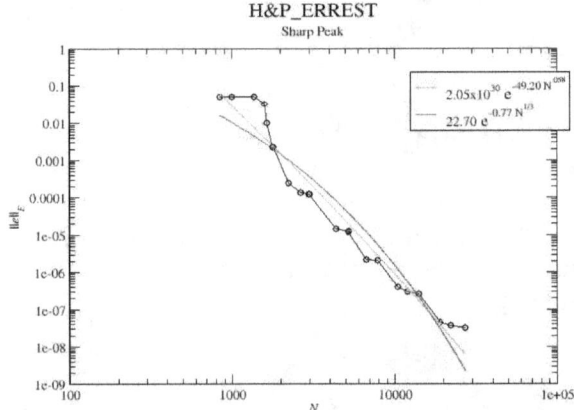

Figure 288: Log-Log plot of the convergence of the COEF-DECAY strategy with the sharp peak problem.

Figure 290: Log-Log plot of the convergence of the H&P-ERREST strategy with the sharp peak problem.

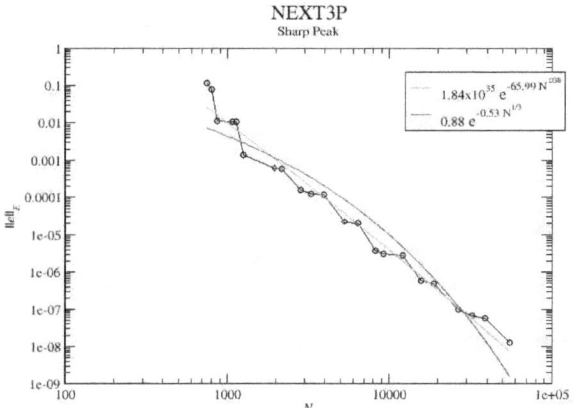

Figure 291: Log-Log plot of the convergence of the NEXT3P strategy with the sharp peak problem.

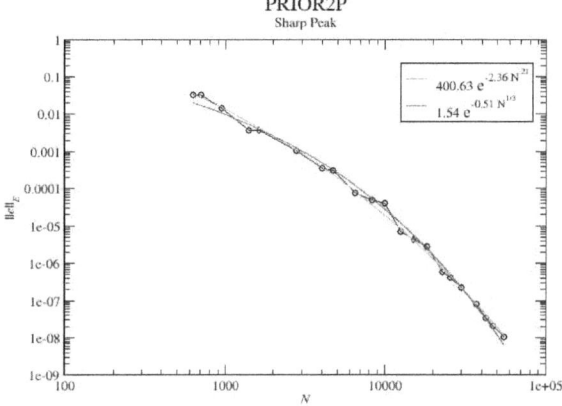

Figure 293: Log-Log plot of the convergence of the PRIOR2P strategy with the sharp peak problem.

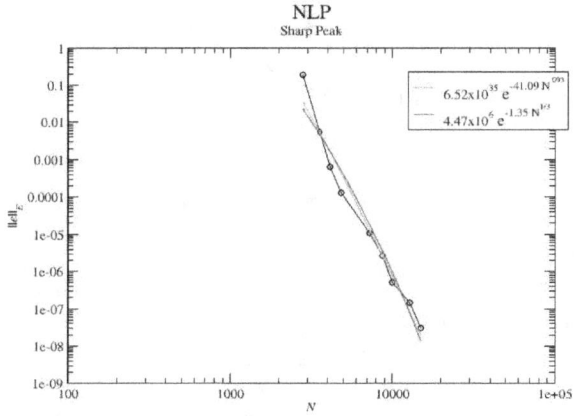

Figure 292: Log-Log plot of the convergence of the NLP strategy with the sharp peak problem.

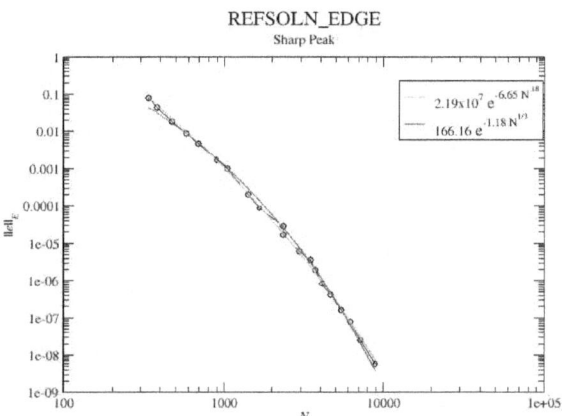

Figure 294: Log-Log plot of the convergence of the REFSOLN_EDGE strategy with the sharp peak problem.

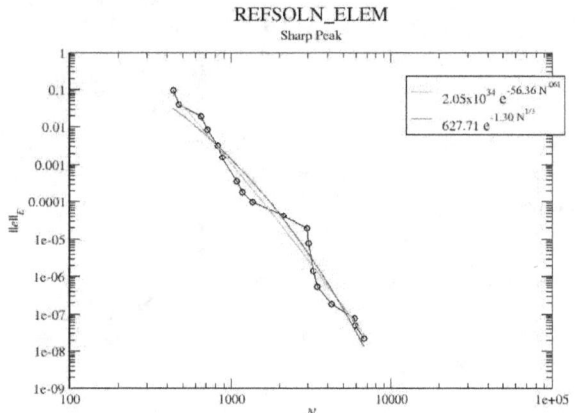
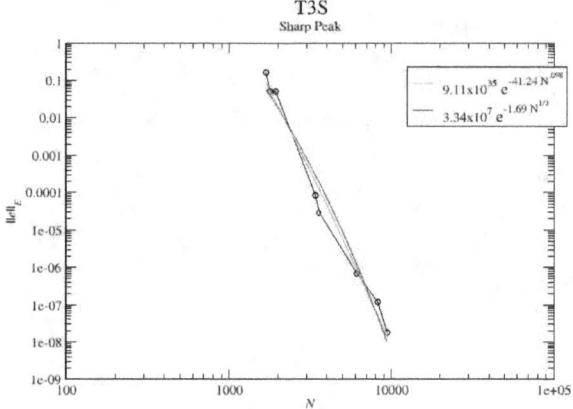

Figure 295: Log-Log plot of the convergence of the REFSOLN_ELEM strategy with the sharp peak problem.

Figure 297: Log-Log plot of the convergence of the T3S strategy with the sharp peak problem.

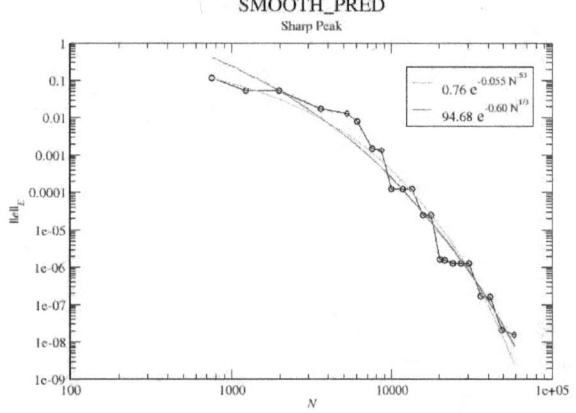
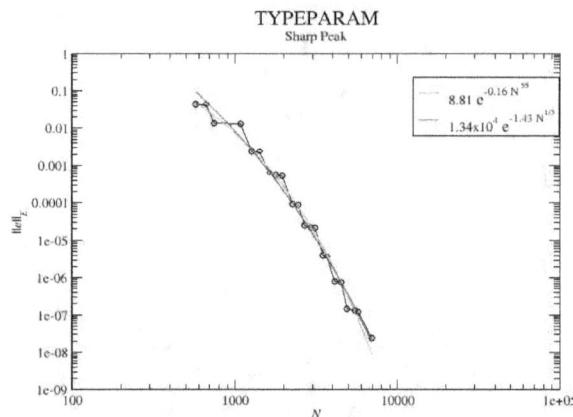

Figure 296: Log-Log plot of the convergence of the SMOOTH_PRED strategy with the sharp peak problem.

Figure 298: Log-Log plot of the convergence of the TYPEPARAM strategy with the sharp peak problem.

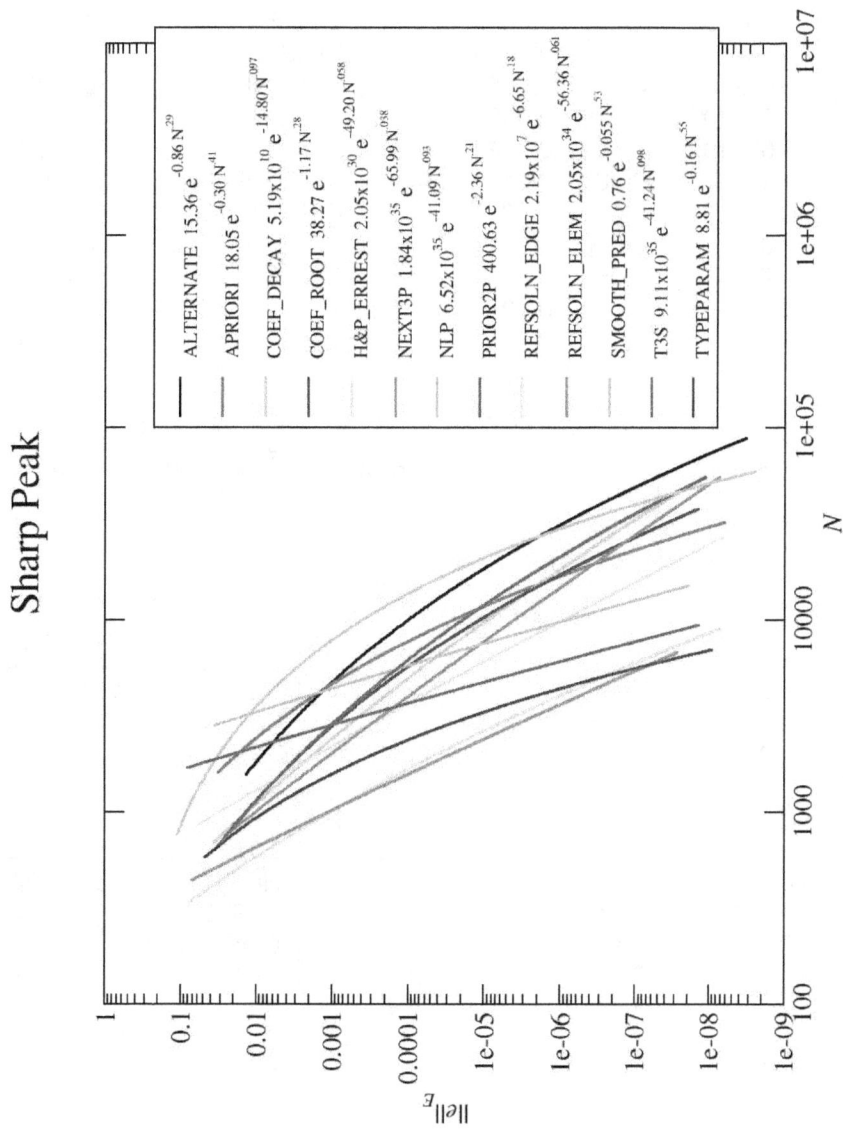

Figure 299: Log-Log plot of the convergence of all strategies with the sharp peak problem.

strategy	A	B	C
ALTERNATE	15.36	-0.86	0.29
APRIORI	18.05	-0.30	0.41
COEFDECAY	5.19×10^{10}	-14.80	0.097
COEFROOT	38.27	-1.17	0.28
H&PERREST	2.05×10^{30}	-49.20	0.058
NEXT3P	1.84×10^{35}	-65.99	0.038
NLP	6.52×10^{35}	-41.09	0.093
PRIOR2P	400.63	-2.36	0.21
REFSOLNEDGE	2.19×10^{7}	-6.65	0.18
REFSOLNELEM	2.05×10^{34}	-56.36	0.061
SMOOTHPRED	0.76	-0.055	0.53
T3S	9.11×10^{35}	-41.24	0.098
TYPEPARAM	8.81	-0.16	0.55

Table 37: Parameters of the least squares fit for $||e_{hp}||_E = Ae^{BN_{dof}^C}$ for the sharp peak problem.

strategy	A	B
ALTERNATE	2.29	-0.46
APRIORI	370.62	-0.77
COEFDECAY	1.19	-0.52
COEFROOT	4.32	-0.59
H&PERREST	22.70	-0.77
NEXT3P	0.88	-0.53
NLP	4.47×10^{6}	-1.35
PRIOR2P	1.54	-0.51
REFSOLNEDGE	166.16	-1.18
REFSOLNELEM	627.71	-1.30
SMOOTHPRED	94.68	-0.60
T3S	3.34×10^{7}	-1.69
TYPEPARAM	1.34×10^{4}	-1.43

Table 39: Parameters of the least squares fit for $||e_{hp}||_E = Ae^{BN_{dof}^{1/3}}$ for the sharp peak problem.

strategy	factor
REFSOLNEDGE	1.00
REFSOLNELEM	1.12
TYPEPARAM	1.59
NEXT3P	1.73
COEFDECAY	1.90
PRIOR2P	1.96
COEFROOT	1.98
H&PERREST	2.23
ALTERNATE	3.13
T3S	3.73
APRIORI	4.15
NLP	5.67
SMOOTHPRED	6.06

Table 38: Factor by which N is larger than the best strategy for the sharp peak problem at low accuracy, 1.0×10^{-2}.

strategy	factor
REFSOLNELEM	1.00
REFSOLNEDGE	1.11
TYPEPARAM	1.21
T3S	1.67
NLP	2.70
H&PERREST	2.71
NEXT3P	4.06
APRIORI	4.66
COEFROOT	4.75
COEFDECAY	5.06
PRIOR2P	5.89
SMOOTHPRED	8.22
ALTERNATE	8.62

Table 40: Factor by which N is larger than the best strategy for the sharp peak problem at high accuracy, 1.0×10^{-6}.

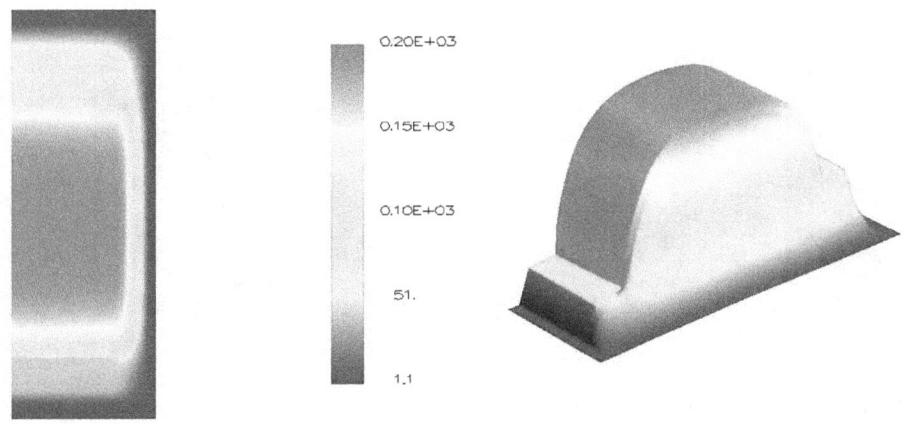

Figure 300: The solution of the battery problem.

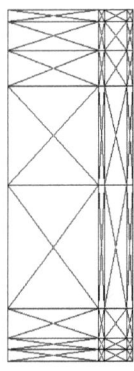

Figure 301: The initial grid for the battery problem.

5.11 Battery

The battery problem is from a model of heat conduction in a battery with nonhomogeneous materials. It has piecewise constant coefficients and right hand side, and mixed boundary conditions on a rectangular domain. The initial grid, shown in Figure 301, is aligned to the discontinuities in the data. The solution has several point singularities in the interior of the domain where three or more materials meet. See [21] for further details. The exact solution of this problem is not known, so the error estimate η (Section 3) is used for the convergence results instead of the error. For the grid images, we used $\tau=10^{-2}$ for most strategies, and $\tau=10^{-1}$ for COEF-ROOT, REFSOLNEDGE and TYPEPARAM. The APRIORI strategy refines by h if the element touches any of the singularities, and by p otherwise.

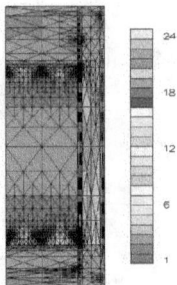
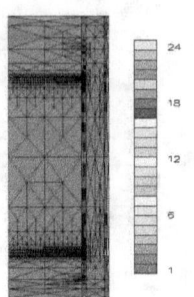

Figure 302: Example grid for the ALTERNATE strategy with the battery problem.

Figure 305: Example grid for the COEFROOT strategy with the battery problem.

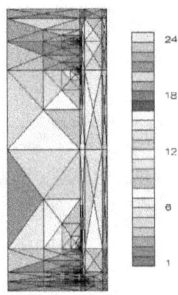
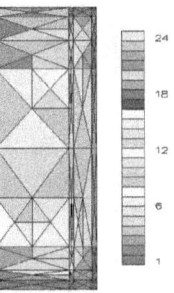

Figure 303: Example grid for the APRIORI strategy with the battery problem.

Figure 306: Example grid for the H&PERREST strategy with the battery problem.

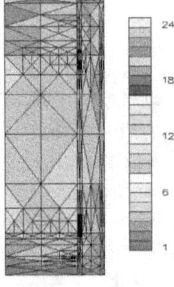
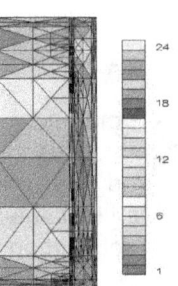

Figure 304: Example grid for the COEFDECAY strategy with the battery problem.

Figure 307: Example grid for the NEXT3P strategy with the battery problem.

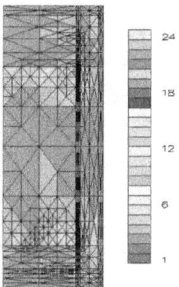

Figure 308: Example grid for the NLP strategy with the battery problem.

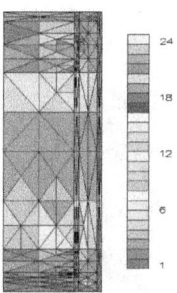

Figure 311: Example grid for the REFSOLNELEM strategy with the battery problem.

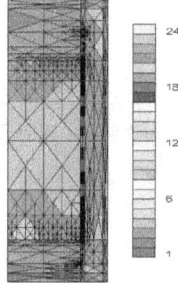

Figure 309: Example grid for the PRIOR2P strategy with the battery problem.

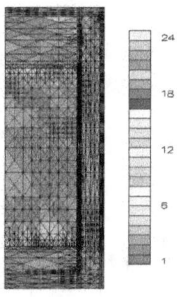

Figure 312: Example grid for the SMOOTHPRED strategy with the battery problem.

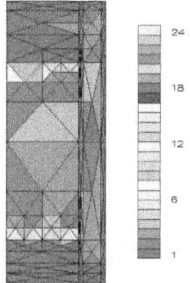

Figure 310: Example grid for the REFSOLNEDGE strategy with the battery problem.

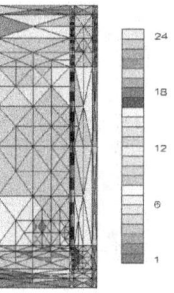

Figure 313: Example grid for the T3S strategy with the battery problem.

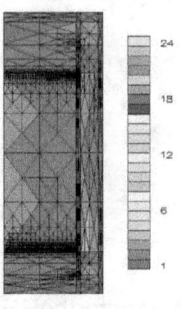

Figure 314: Example grid for the TYPEPARAM strategy with the battery problem.

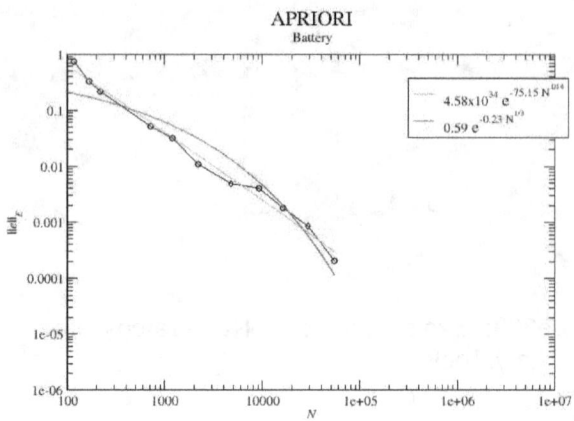

Figure 316: Log-Log plot of the convergence of the APRIORI strategy with the battery problem.

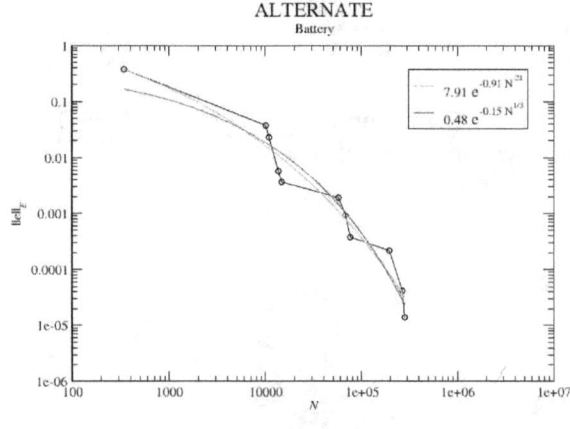

Figure 315: Log-Log plot of the convergence of the ALTERNATE strategy with the battery problem.

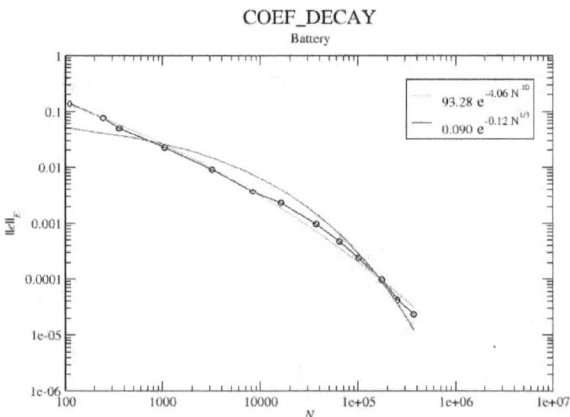

Figure 317: Log-Log plot of the convergence of the COEF-DECAY strategy with the battery problem.

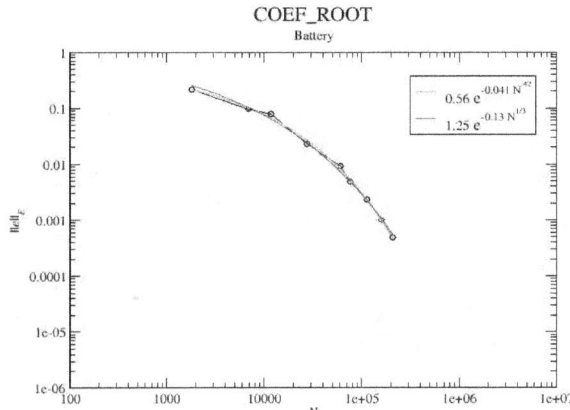

Figure 318: Log-Log plot of the convergence of the COEF-ROOT strategy with the battery problem.

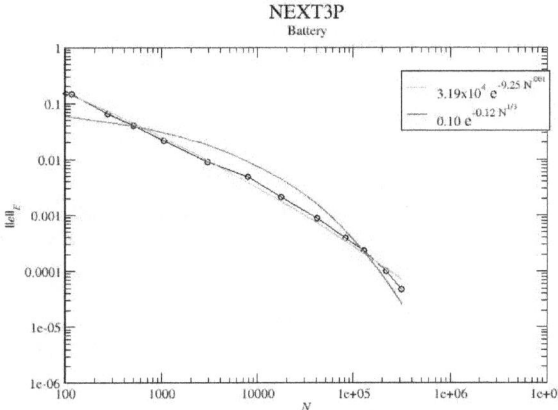

Figure 320: Log-Log plot of the convergence of the NEXT3P strategy with the battery problem.

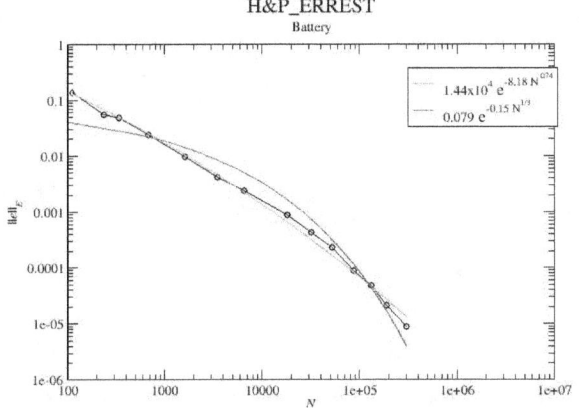

Figure 319: Log-Log plot of the convergence of the H&P-ERREST strategy with the battery problem.

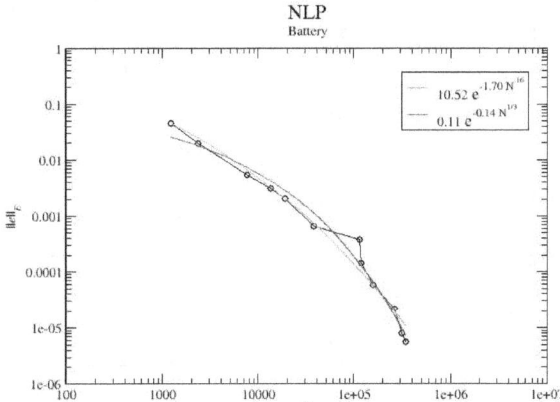

Figure 321: Log-Log plot of the convergence of the NLP strategy with the battery problem.

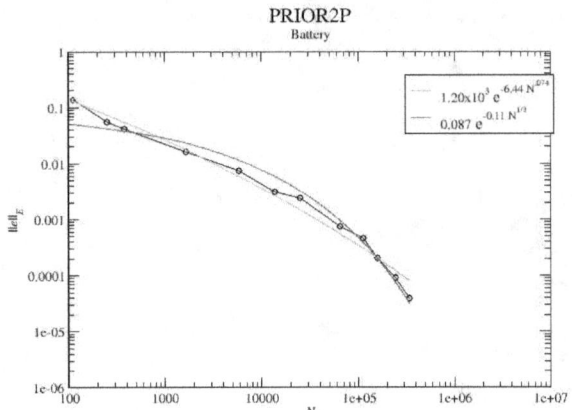

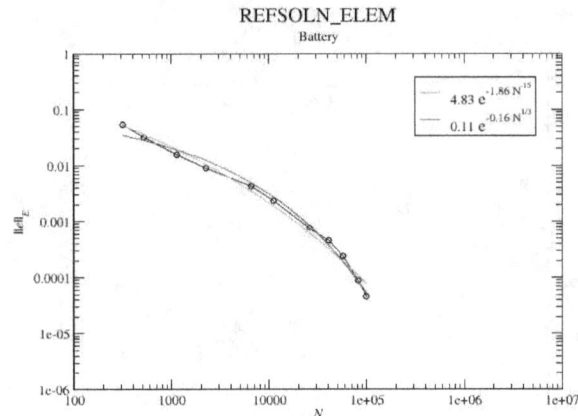

Figure 322: Log-Log plot of the convergence of the PRIOR2P strategy with the battery problem.

Figure 324: Log-Log plot of the convergence of the REFSOLN-ELEM strategy with the battery problem.

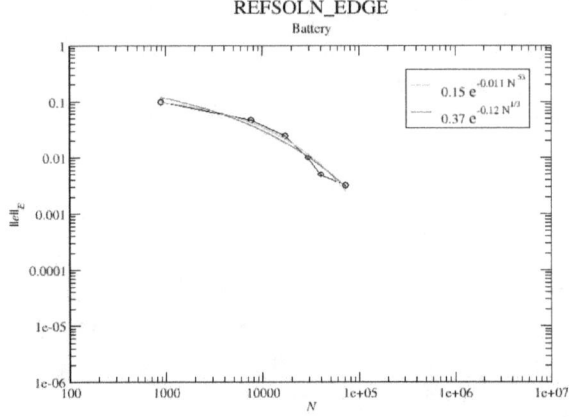

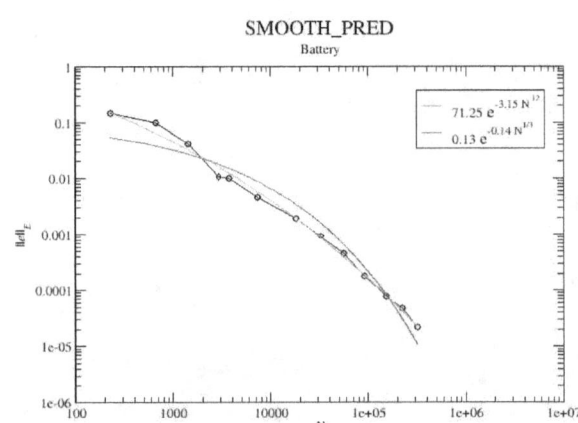

Figure 323: Log-Log plot of the convergence of the REFSOLN-EDGE strategy with the battery problem.

Figure 325: Log-Log plot of the convergence of the SMOOTH-PRED strategy with the battery problem.

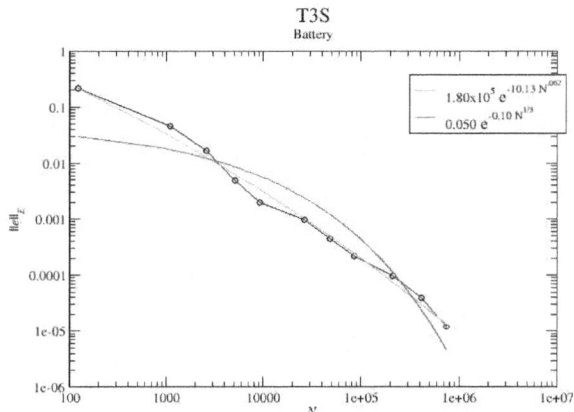

Figure 326: Log-Log plot of the convergence of the T3S strategy with the battery problem.

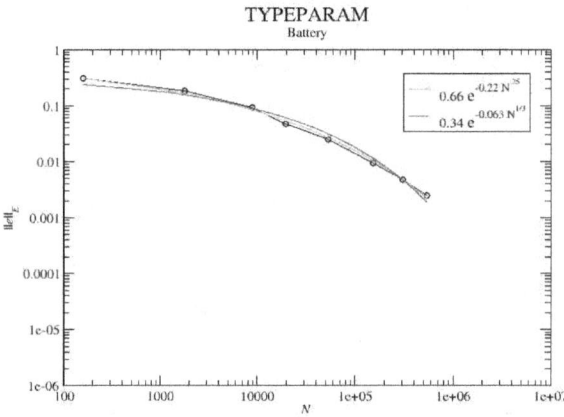

Figure 327: Log-Log plot of the convergence of the TYPEPARAM strategy with the battery problem.

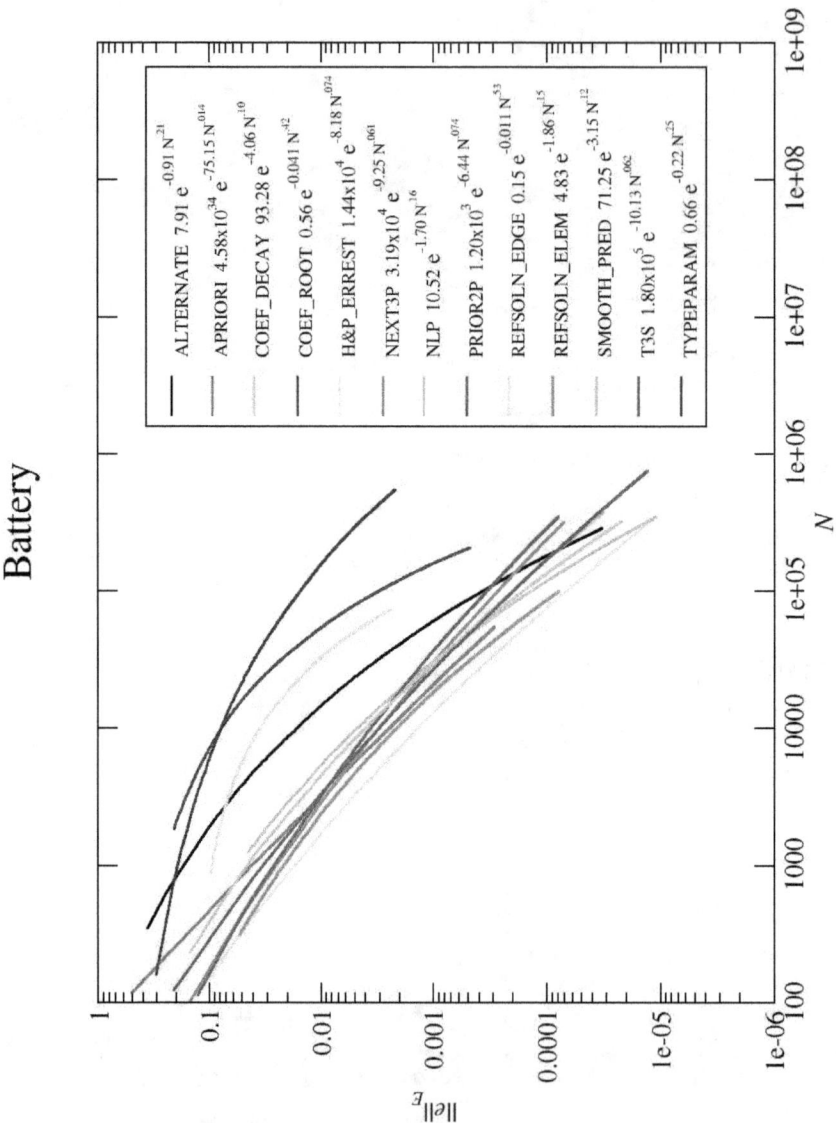

Figure 328: Log-Log plot of the convergence of all strategies with the battery problem.

strategy	A	B	C
ALTERNATE	7.91	-0.91	0.21
APRIORI	4.58x10³⁴	-75.15	0.014
COEFDECAY	93.28	-4.06	0.10
COEFROOT	0.56	-0.041	0.42
H&PERREST	1.44x10⁴	-8.18	0.074
NEXT3P	3.19x10⁴	-9.25	0.061
NLP	10.52	-1.70	0.16
PRIOR2P	1.20x10³	-6.44	0.074
REFSOLNEDGE	0.15	-0.011	0.53
REFSOLNELEM	4.83	-1.86	0.15
SMOOTHPRED	71.25	-3.15	0.12
T3S	1.80x10⁶	-10.13	0.062
TYPEPARAM	0.66	-0.22	0.25

Table 41: Parameters of the least squares fit for $\|e_{hp}\|_E = Ae^{BN_{dof}^C}$ for the battery problem.

strategy	A	B
ALTERNATE	0.48	-0.15
APRIORI	0.59	-0.23
COEFDECAY	0.090	-0.12
COEFROOT	1.25	-0.13
H&PERREST	0.079	-0.15
NEXT3P	0.10	-0.12
NLP	0.11	-0.14
PRIOR2P	0.087	-0.11
REFSOLNEDGE	0.37	-0.12
REFSOLNELEM	0.11	-0.16
SMOOTHPRED	0.13	-0.14
T3S	0.050	-0.10
TYPEPARAM	0.34	-0.063

Table 43: Parameters of the least squares fit for $\|e_{hp}\|_E = Ae^{BN_{dof}^{1/3}}$ for the battery problem.

strategy	factor
H&PERREST	1.00
REFSOLNELEM	1.39
NEXT3P	1.68
COEFDECAY	1.78
APRIORI	1.89
PRIOR2P	1.89
T3S	1.94
SMOOTHPRED	2.66
NLP	3.17
ALTERNATE	8.56
REFSOLNEDGE	19.22
COEFROOT	31.82
TYPEPARAM	92.09

Table 42: Factor by which N is larger than the best strategy for the battery problem at low accuracy, 1.0×10^{-2}.

strategy	factor
H&PERREST	1.00
REFSOLNELEM	1.34
APRIORI	1.59
NLP	2.10
T3S	2.11
SMOOTHPRED	2.12
COEFDECAY	2.19
NEXT3P	2.56
PRIOR2P	3.12
ALTERNATE	3.81
REFSOLNEDGE	5.97
COEFROOT	8.97
TYPEPARAM	61.73

Table 44: Factor by which N is larger than the best strategy for the battery problem at high accuracy, 5.0×10^{-4}.

Figure 329: The solution of the mild boundary layer problem.

5.12 Boundary Layer, Mild

The boundary layer problem is a convection-diffusion equation with first order terms and Dirichlet boundary conditions on $(-1,1) \times (-1,1)$. The solution is

$$(1 - e^{-(1-x)/\varrho})(1 - e^{-(1-y)/\varrho})\cos(\pi(x+y))$$

where ϱ controls the strength of the boundary layer. In the easy form of this problem we use $\varrho = 10^{-1}$. For the grid images, $\tau = 10^{-4}$. In the APRIORI strategy we refine by h if the element touches either of the boundaries with a boundary layer, and by p otherwise.

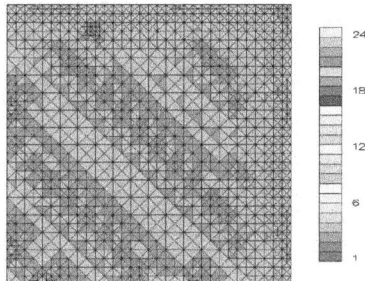

Figure 330: Example grid for the ALTERNATE strategy with the mild boundary layer problem.

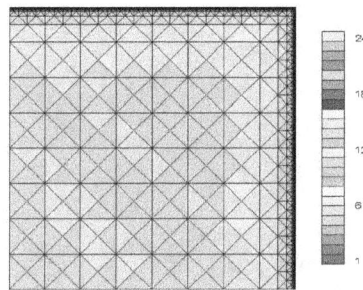

Figure 331: Example grid for the APRIORI strategy with the mild boundary layer problem.

Figure 332: Example grid for the COEFDECAY strategy with the mild boundary layer problem.

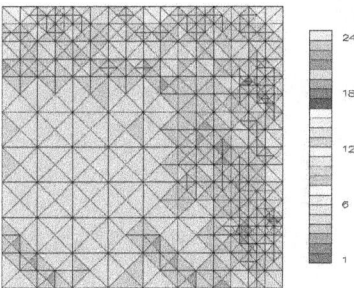

Figure 333: Example grid for the COEFROOT strategy with the mild boundary layer problem.

Figure 334: Example grid for the H&PERREST strategy with the mild boundary layer problem.

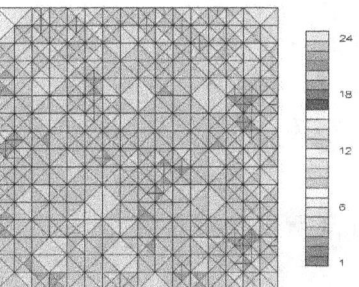

Figure 335: Example grid for the NEXT3P strategy with the mild boundary layer problem.

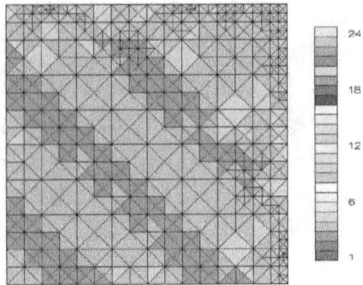

Figure 336: Example grid for the NLP strategy with the mild boundary layer problem.

Figure 339: Example grid for the REFSOLNELEM strategy with the mild boundary layer problem.

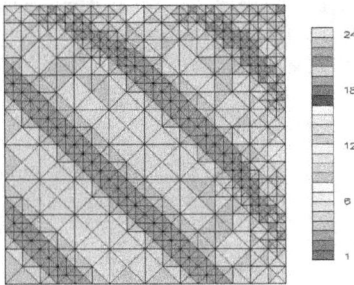

Figure 337: Example grid for the PRIOR2P strategy with the mild boundary layer problem.

Figure 340: Example grid for the SMOOTHPRED strategy with the mild boundary layer problem.

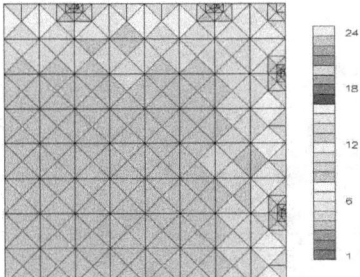

Figure 338: Example grid for the REFSOLNEDGE strategy with the mild boundary layer problem.

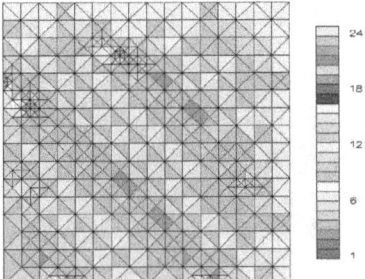

Figure 341: Example grid for the T3S strategy with the mild boundary layer problem.

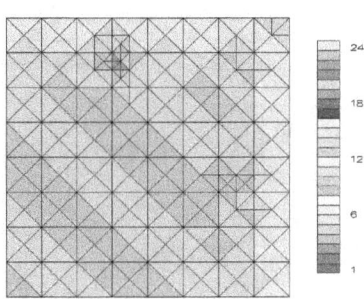

Figure 342: Example grid for the TYPEPARAM strategy with the mild boundary layer problem.

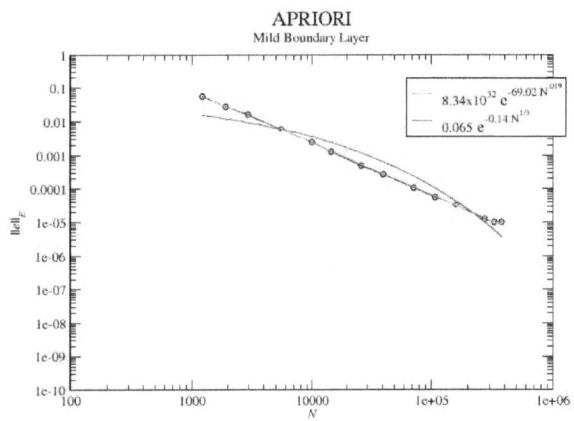

Figure 344: Log-Logplot of the convergence of the APRIORI strategy with the mild boundary layer problem.

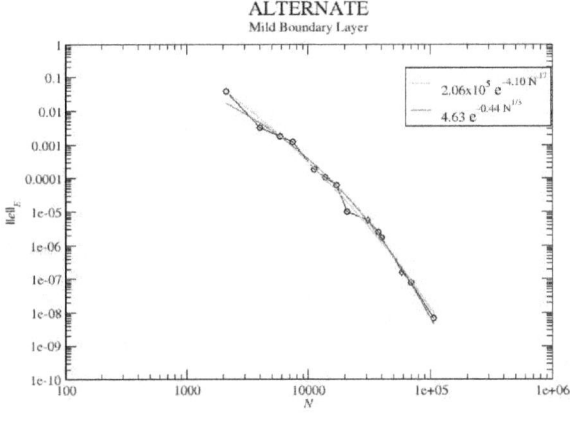

Figure 343: Log-Logplot of the convergence of the ALTERNATE strategy with the mild boundary layer problem.

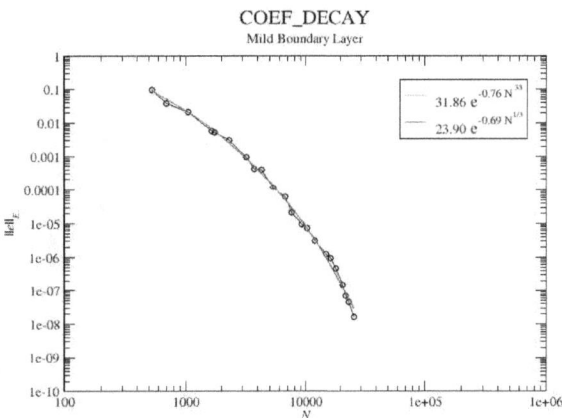

Figure 345: Log-Logplot of the convergence of the COEF-DECAY strategy with the mild boundary layer problem.

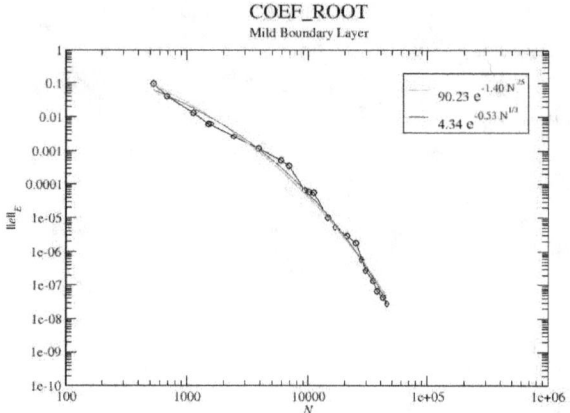

Figure 346: Log-Logplot of the convergence of the COEF-ROOT strategy with the mild boundary layer problem.

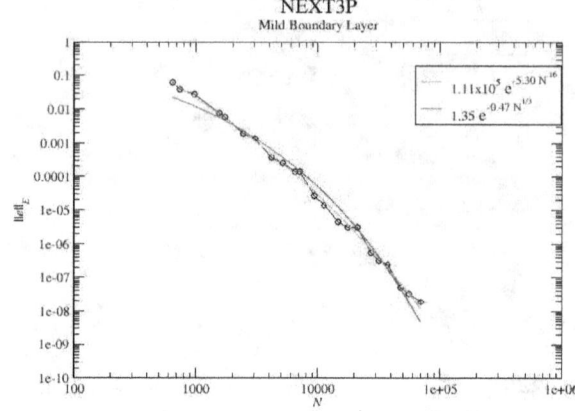

Figure 348: Log-Logplot of the convergence of the NEXT3P strategy with the mild boundary layer problem.

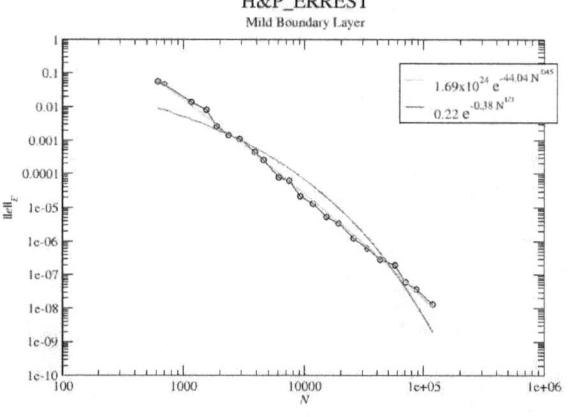

Figure 347: Log-Logplot of the convergence of the H&P-ERREST strategy with the mild boundary layer problem.

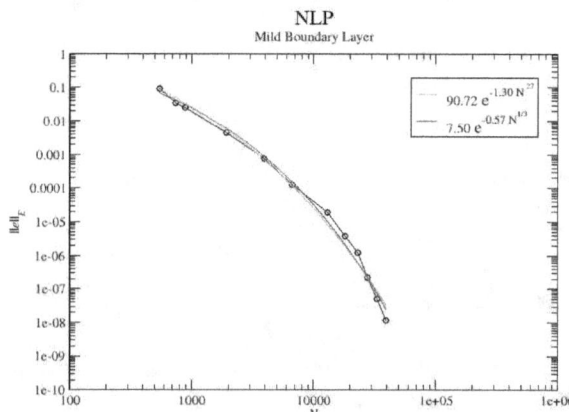

Figure 349: Log-Logplot of the convergence of the NLP strategy with the mild boundary layer problem.

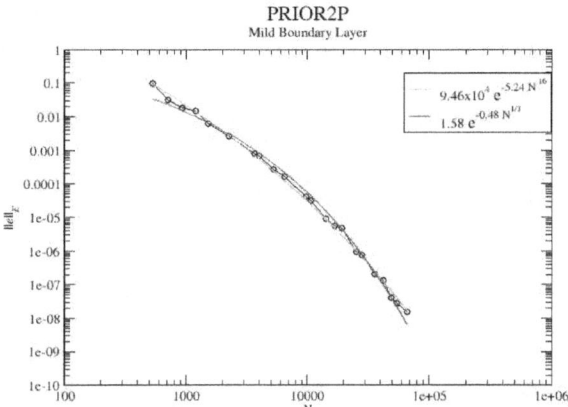
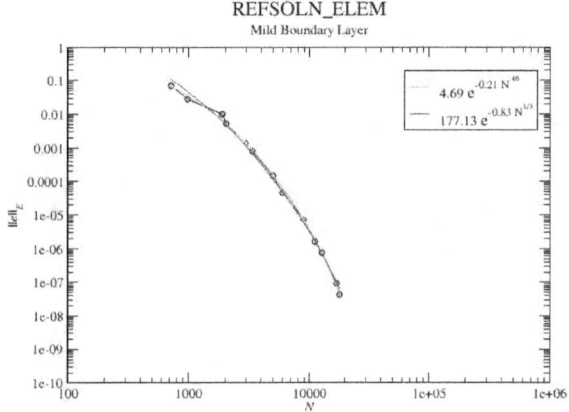

Figure 350: Log-Log plot of the convergence of the PRIOR2P strategy with the mild boundary layer problem.

Figure 352: Log-Log plot of the convergence of the REFSOLN-ELEM strategy with the mild boundary layer problem.

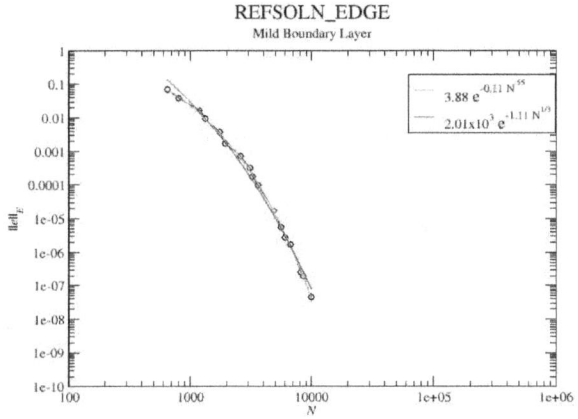
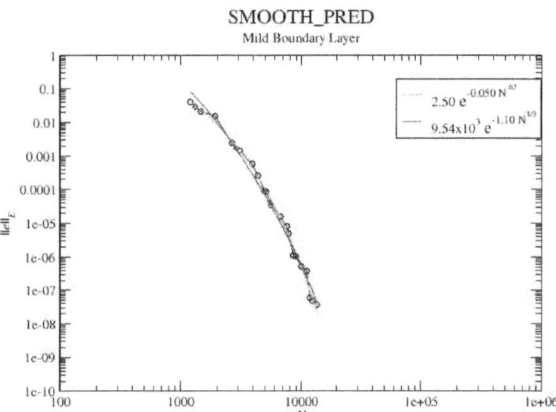

Figure 351: Log-Log plot of the convergence of the REFSOLN-EDGE strategy with the mild boundary layer problem.

Figure 353: Log-Log plot of the convergence of the SMOOTH-PRED strategy with the mild boundary layer problem.

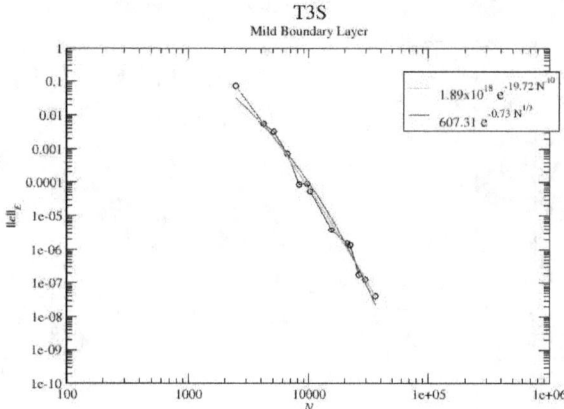

Figure 354: Log-Log plot of the convergence of the T3S strategy with the mild boundary layer problem.

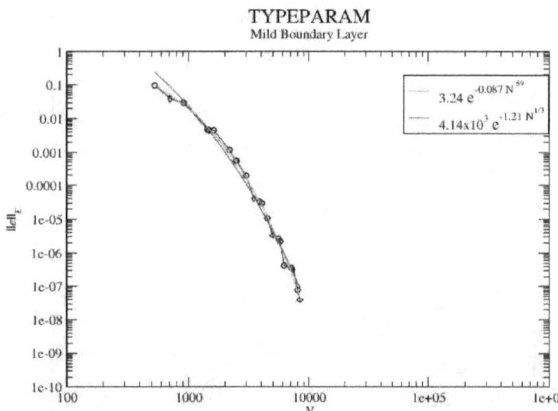

Figure 355: Log-Log plot of the convergence of the TYPEPARAM strategy with the mild boundary layer problem.

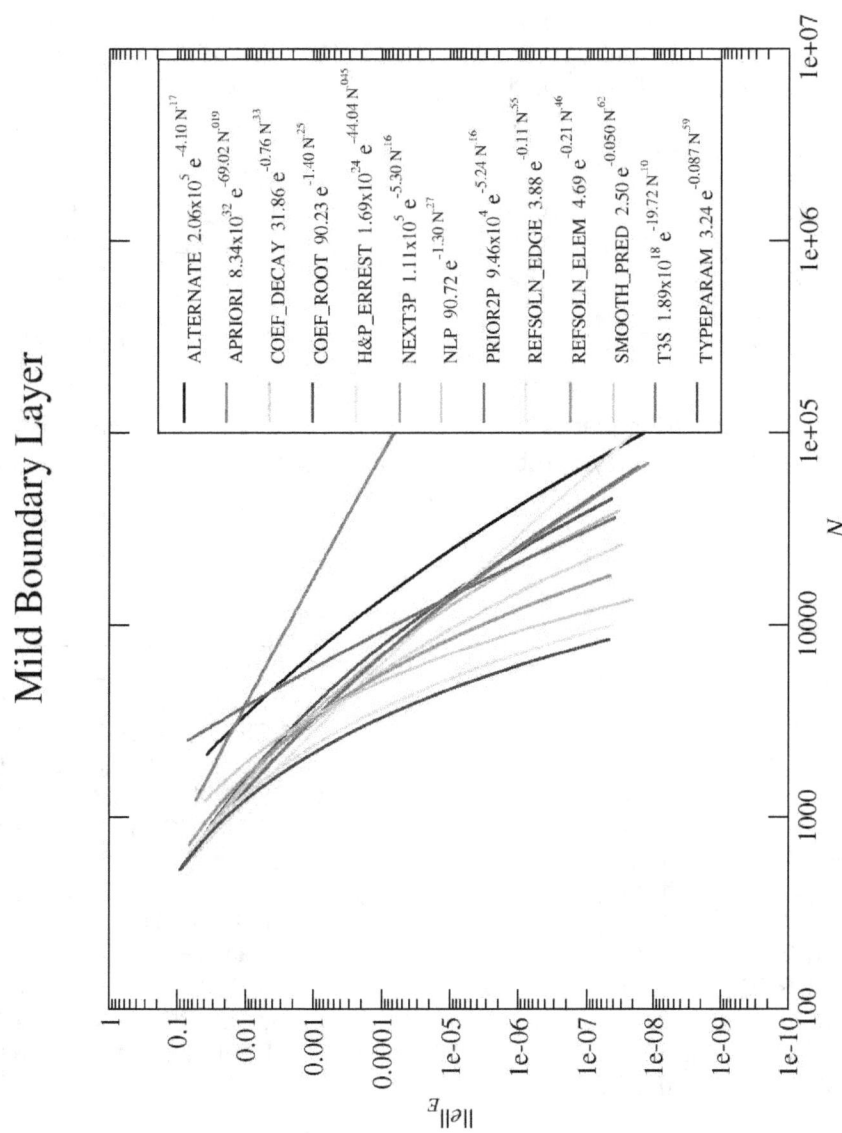

Figure 356: Log-Log plot of the convergence of all strategies with the mild boundary layer problem.

strategy	A	B	C
ALTERNATE	2.06×10^5	-4.10	0.17
APRIORI	8.34×10^{32}	-69.02	0.019
COEFDECAY	31.86	-0.76	0.33
COEFROOT	90.23	-1.40	0.25
H&PERREST	1.69×10^{24}	-44.04	0.045
NEXT3P	1.11×10^5	-5.30	0.16
NLP	90.72	-1.30	0.27
PRIOR2P	9.46×10^4	-5.24	0.16
REFSOLNEDGE	3.88	-0.11	0.55
REFSOLNELEM	4.69	-0.21	0.46
SMOOTHPRED	2.50	-0.050	0.62
T3S	1.89×10^8	-19.72	0.10
TYPEPARAM	3.24	-0.087	0.59

Table 45: Parameters of the least squares fit for $\|e_{hp}\|_E = A e^{BN_{dof}^C}$ for the mild boundary layer problem.

strategy	A	B
ALTERNATE	4.63	-0.44
APRIORI	0.065	-0.14
COEFDECAY	23.90	-0.69
COEFROOT	4.34	-0.53
H&PERREST	0.22	-0.38
NEXT3P	1.35	-0.47
NLP	7.50	-0.57
PRIOR2P	1.58	-0.48
REFSOLNEDGE	2.01×10^3	-1.11
REFSOLNELEM	177.13	-0.83
SMOOTHPRED	9.54×10^3	-1.10
T3S	607.31	-0.73
TYPEPARAM	4.14×10^3	-1.21

Table 47: Parameters of the least squares fit for $\|e_{hp}\|_E = A e^{BN_{dof}^{1/3}}$ for the mild boundary layer problem.

strategy	factor
H&PERREST	1.00
TYPEPARAM	1.02
REFSOLNEDGE	1.08
NEXT3P	1.14
PRIOR2P	1.14
COEFDECAY	1.18
NLP	1.25
COEFROOT	1.30
REFSOLNELEM	1.35
SMOOTHPRED	1.59
ALTERNATE	2.82
T3S	3.13
APRIORI	3.20

Table 46: Factor by which N is larger than the best strategy for the mild boundary layer problem at low accuracy, 1.0×10^{-2}.

strategy	factor
TYPEPARAM	1.00
REFSOLNEDGE	1.15
SMOOTHPRED	1.51
REFSOLNELEM	1.96
COEFDECAY	2.41
NLP	3.38
T3S	3.42
COEFROOT	3.97
NEXT3P	4.05
PRIOR2P	4.14
H&PERREST	4.75
ALTERNATE	6.80
APRIORI	219.71

Table 48: Factor by which N is larger than the best strategy for the mild boundary layer problem at high accuracy, 1.0×10^{-6}.

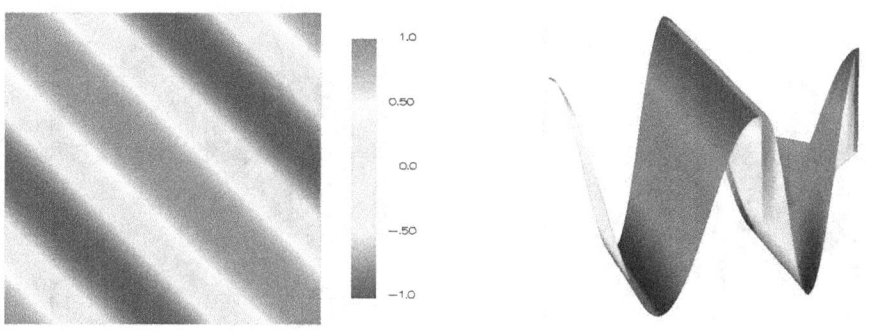

Figure 357: The solution of the strong boundary layer problem.

5.13 Boundary Layer, Strong

For the hard version of the boundary layer problem (Section 5.12) we use $\varrho = 10^{-3}$. For the grid images, $\tau = 10^{-1}$. In the APRIORI strategy we refine by h if the element touches either of the boundaries with a boundary layer, and by p otherwise.

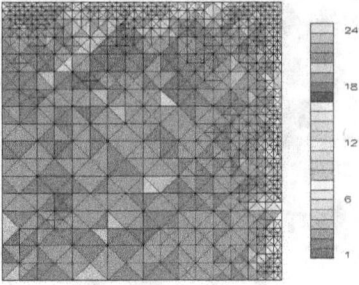

Figure 358: Example grid for the ALTERNATE strategy with the strong boundary layer problem.

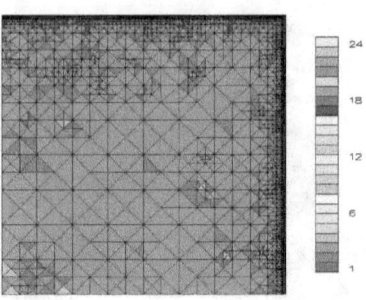

Figure 361: Example grid for the COEFROOT strategy with the strong boundary layer problem.

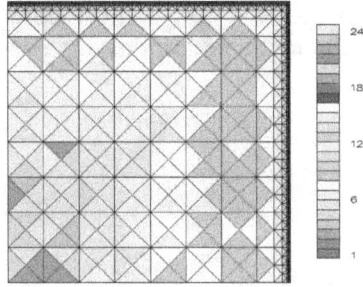

Figure 359: Example grid for the APRIORI strategy with the strong boundary layer problem.

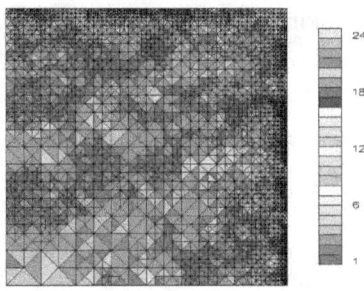

Figure 362: Example grid for the H&PERREST strategy with the strong boundary layer problem.

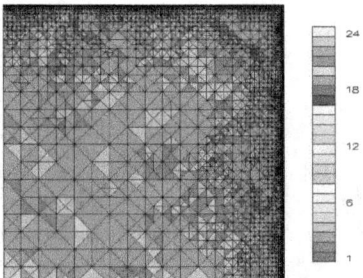

Figure 360: Example grid for the COEFDECAY strategy with the strong boundary layer problem.

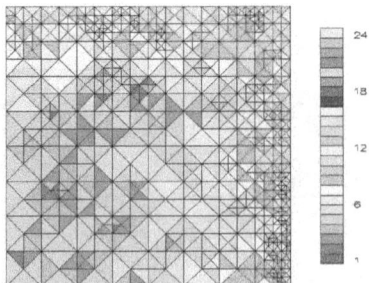

Figure 363: Example grid for the NEXT3P strategy with the strong boundary layer problem.

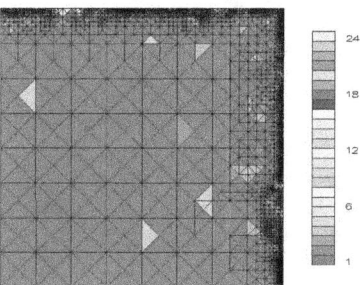

Figure 364: Example grid for the NLP strategy with the strong boundary layer problem.

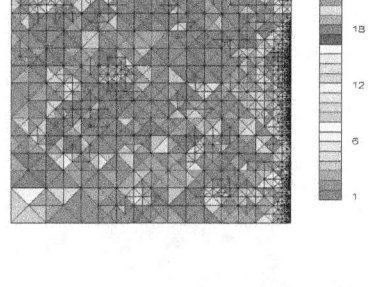

Figure 367: Example grid for the REFSOLNELEM strategy with the strong boundary layer problem.

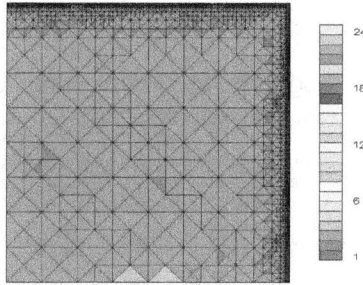

Figure 365: Example grid for the PRIOR2P strategy with the strong boundary layer problem.

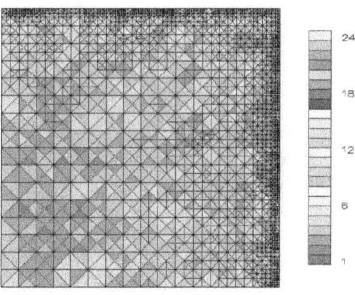

Figure 368: Example grid for the SMOOTHPRED strategy with the strong boundary layer problem.

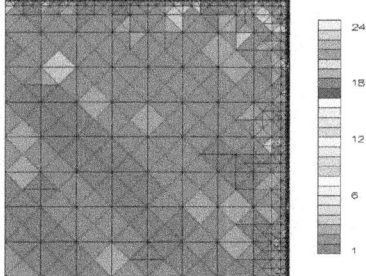

Figure 366: Example grid for the REFSOLNEDGE strategy with the strong boundary layer problem.

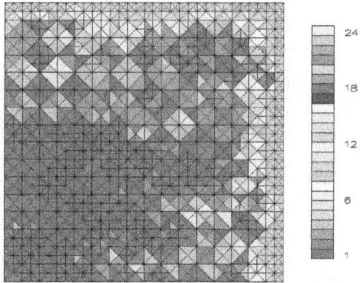

Figure 369: Example grid for the T3S strategy with the strong boundary layer problem.

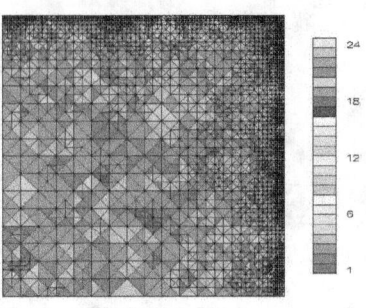

Figure 370: Example grid for the TYPEPARAM strategy with the strong boundary layer problem.

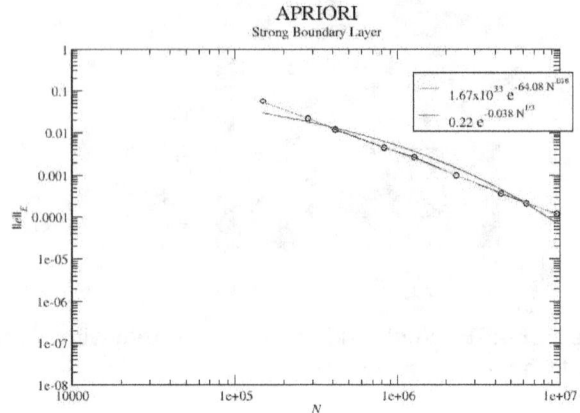

Figure 372: Log-Log plot of the convergence of the APRIORI strategy with the strong boundary layer problem.

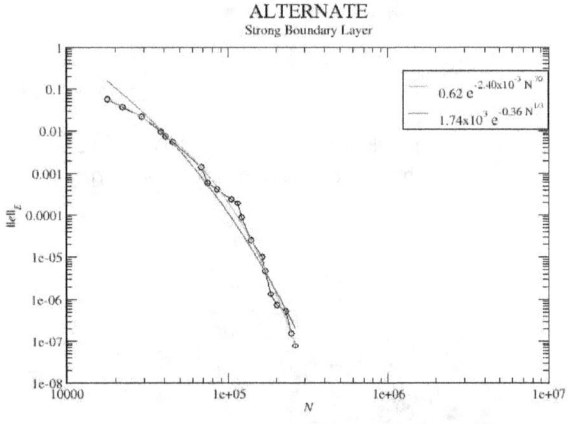

Figure 371: Log-Log plot of the convergence of the ALTERNATE strategy with the strong boundary layer problem.

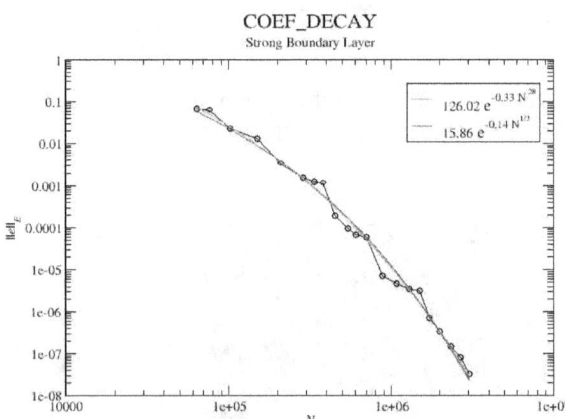

Figure 373: Log-Log plot of the convergence of the COEF-DECAY strategy with the strong boundary layer problem.

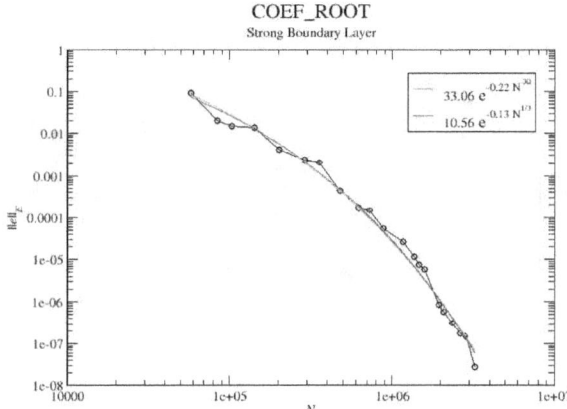

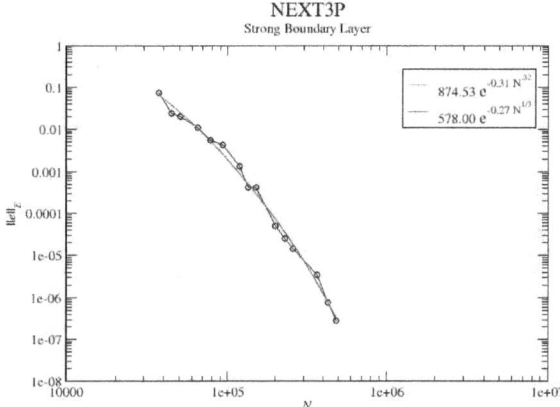

Figure 374: Log-Log plot of the convergence of the COEF-ROOT strategy with the strong boundary layer problem.

Figure 376: Log-Log plot of the convergence of the NEXT3P strategy with the strong boundary layer problem.

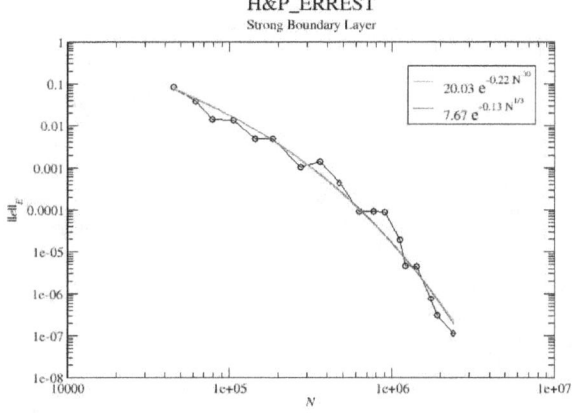

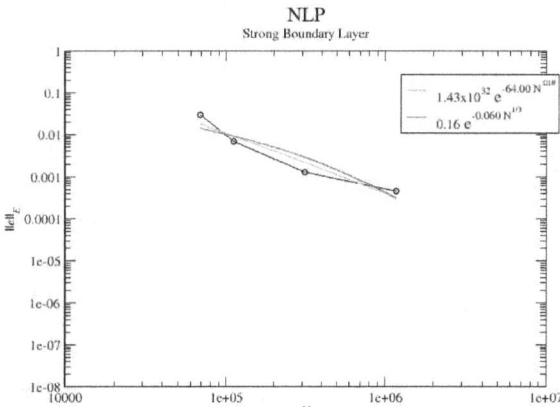

Figure 375: Log-Log plot of the convergence of the H&P-ERREST strategy with the strong boundary layer problem.

Figure 377: Log-Log plot of the convergence of the NLP strategy with the strong boundary layer problem.

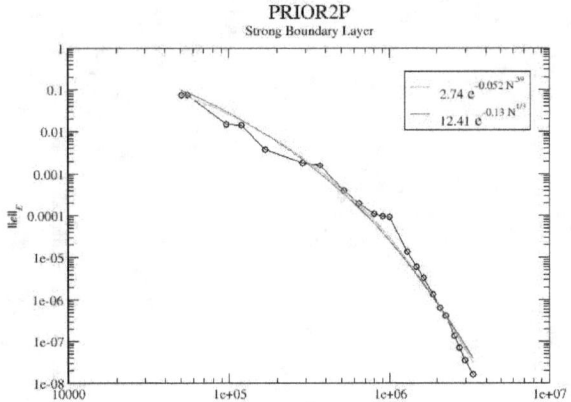

Figure 378: Log-Logplot of the convergence of the PRIOR2P strategy with the strong boundary layer problem.

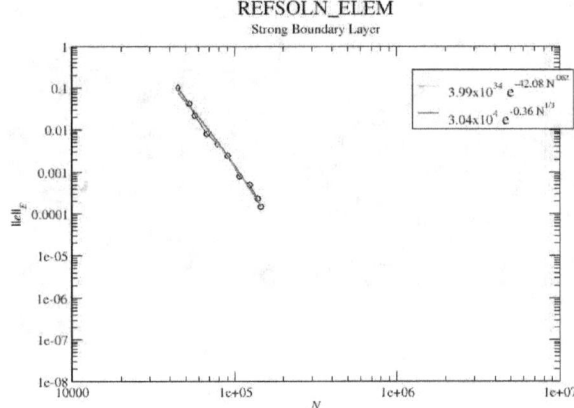

Figure 380: Log-Logplot of the convergence of the REFSOLN-ELEM strategy with the strong boundary layer problem.

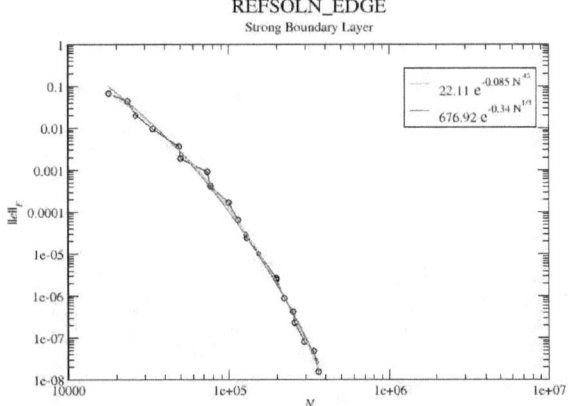

Figure 379: Log-Logplot of the convergence of the REFSOLN-EDGE strategy with the strong boundary layer problem.

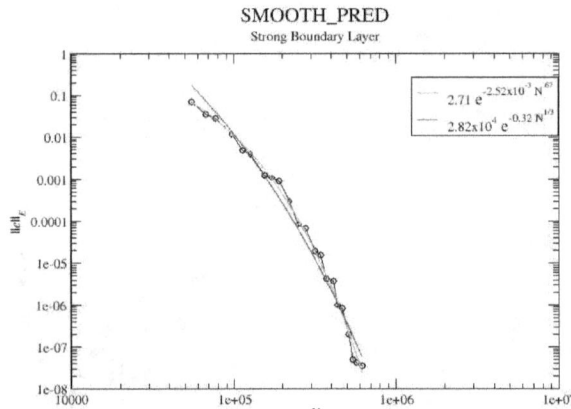

Figure 381: Log-Logplot of the convergence of the SMOOTH-PRED strategy with the strong boundary layer problem.

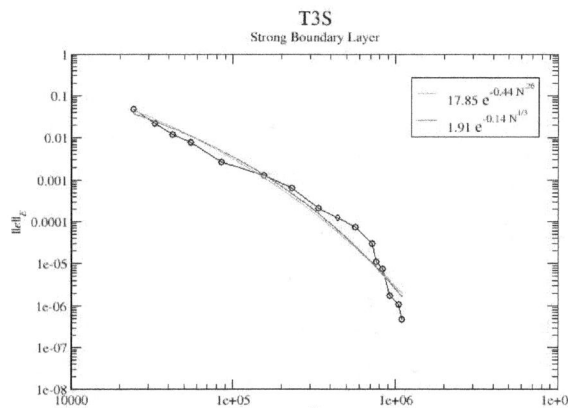

Figure 382: Log-Log plot of the convergence of the T3S strategy with the strong boundary layer problem.

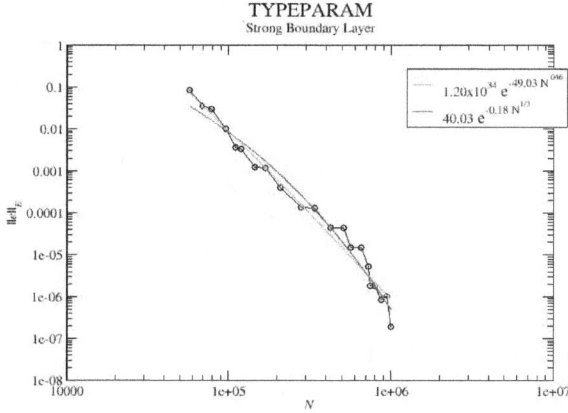

Figure 383: Log-Log plot of the convergence of the TYPEPARAM strategy with the strong boundary layer problem.

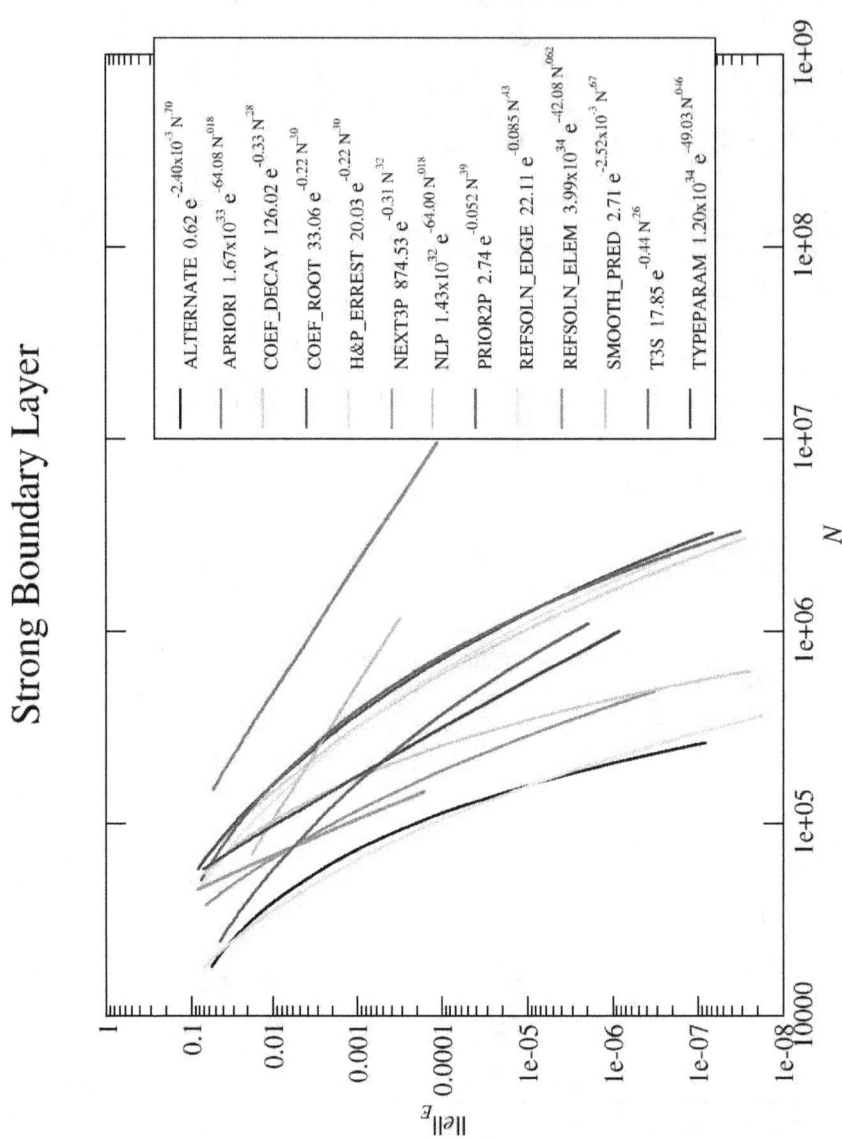

Figure 384: Log-Log plot of the convergence of all strategies with the strong boundary layer problem.

strategy	A	B	C
ALTERNATE	0.62	-2.40×10^{-3}	0.70
APRIORI	1.67×10^{33}	-64.08	0.018
COEFDECAY	126.02	-0.33	0.28
COEFROOT	33.06	-0.22	0.30
H&PERREST	20.03	-0.22	0.30
NEXT3P	874.53	-0.31	0.32
NLP	1.43×10^{32}	-64.00	0.018
PRIOR2P	2.74	-0.052	0.39
REFSOLNEDGE	22.11	-0.085	0.43
REFSOLNELEM	3.99×10^{34}	-42.08	0.062
SMOOTHPRED	2.71	-2.52×10^{-3}	0.67
T3S	17.85	-0.44	0.26
TYPEPARAM	1.20×10^{34}	-49.03	0.046

Table 49: Parameters of the least squares fit for $\|e_{hp}\|_E = Ae^{BN_{dof}^C}$ for the strongboundary layerproblem.

strategy	A	B
ALTERNATE	1.74×10^{3}	-0.36
APRIORI	0.22	-0.038
COEFDECAY	15.86	-0.14
COEFROOT	10.56	-0.13
H&PERREST	7.67	-0.13
NEXT3P	578.00	-0.27
NLP	0.16	-0.060
PRIOR2P	12.41	-0.13
REFSOLNEDGE	676.92	-0.34
REFSOLNELEM	3.04×10^{4}	-0.36
SMOOTHPRED	2.82×10^{4}	-0.32
T3S	1.91	-0.14
TYPEPARAM	40.03	-0.18

Table 51: Parameters of the least squares fit for $\|e_{hp}\|_E = Ae^{BN_{dof}^{1/3}}$ for the strongboundary layerproblem.

strategy	factor
REFSOLNEDGE	1.00
ALTERNATE	1.11
T3S	1.64
NEXT3P	1.87
REFSOLNELEM	1.93
TYPEPARAM	2.76
SMOOTHPRED	2.93
NLP	2.99
H&PERREST	3.71
COEFDECAY	4.17
COEFROOT	4.54
PRIOR2P	4.55
APRIORI	13.73

Table 50: Factor by which N is larger than the best strategy for the strongboundary layer problem at low accuracy, 1.0×10^{-2}.

strategy	factor
ALTERNATE	1.00
REFSOLNEDGE	1.07
REFSOLNELEM	1.75
NEXT3P	2.00
SMOOTHPRED	2.15
TYPEPARAM	4.67
T3S	6.25
COEFDECAY	7.93
H&PERREST	8.78
PRIOR2P	9.40
COEFROOT	9.74
NLP	257.52
APRIORI	926.79

Table 52: Factor by which N is larger than the best strategy for the strongboundary layer problem at high accuracy, 1.0×10^{-6}.

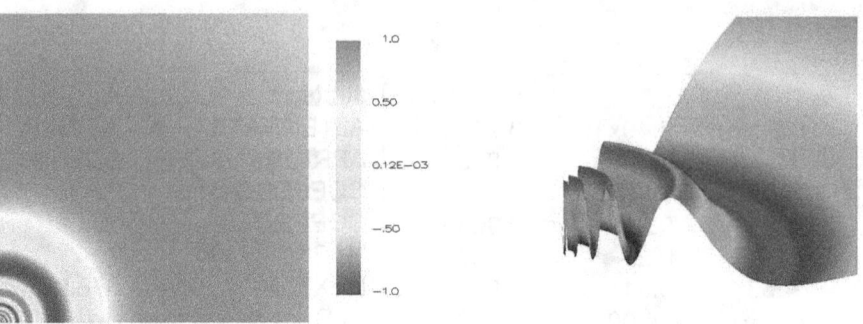

Figure 385: The solution of the mild oscillatory problem.

5.14 Oscillatory, Mild

The oscillatory problem contains several circular waves which get closer together as you approach the origin. The PDE is a Helmholtz equation with Dirichlet boundary conditions on the unit square. The solution is

$$\sin\left(\frac{1}{\alpha+r}\right)$$

where $r = \sqrt{x^2+y^2}$. The number of oscillations, N, is determined by the parameter $\alpha = \frac{1}{N\pi}$. For the easy form of this problem we use N=10.5. $\tau=10^{-3}$ for the grid images. For APRIORI, refine by h if the element touches the origin and by p otherwise. In the perspective view of the solution in Figure 385 we zoomed in on the origin to show the details of the oscillations.

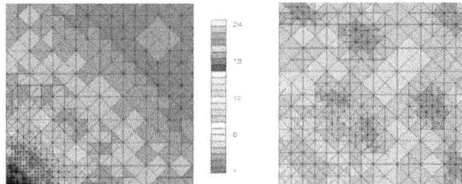

Figure 386: Example grid for the ALTERNATE strategy with the mild oscillatory problem, including details at the origin.

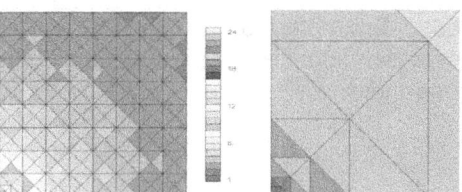

Figure 387: Example grid for the APRIORI strategy with the mild oscillatory problem, including details at the origin.

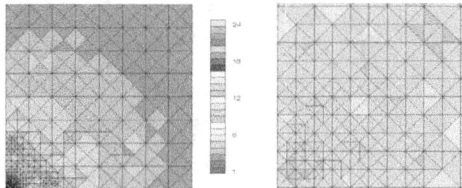

Figure 388: Example grid for the COEFDECAY strategy with the mild oscillatory problem, including details at the origin.

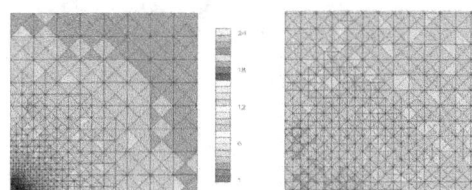

Figure 389: Example grid for the COEFROOT strategy with the mild oscillatory problem, including details at the origin.

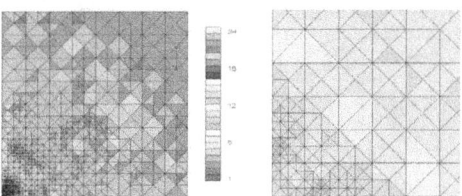

Figure 390: Example grid for the H&PERREST strategy with the mild oscillatory problem, including details at the origin.

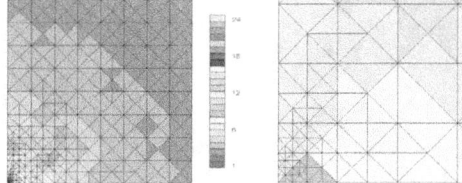

Figure 391: Example grid for the NEXT3P strategy with the mild oscillatory problem, including details at the origin.

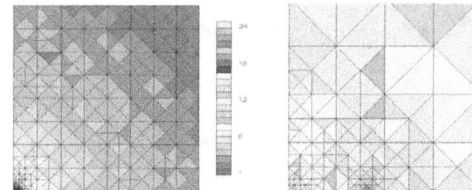

Figure 392: Example grid for the NLP strategy with the mild oscillatory problem, including details at the origin.

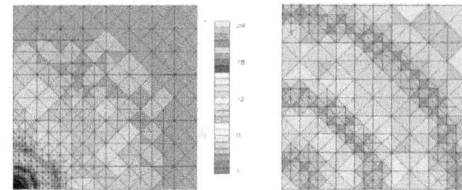

Figure 393: Example grid for the PRIOR2P strategy with the mild oscillatory problem, including details at the origin.

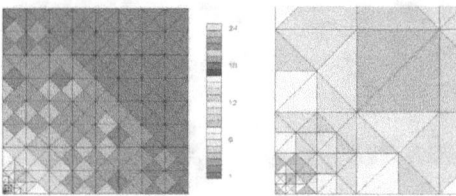

Figure 394: Example grid for the REFSOLNEDGE strategy with the mild oscillatory problem, including details at the origin.

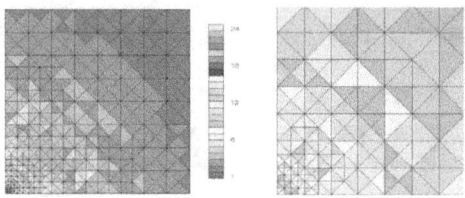

Figure 395: Example grid for the REFSOLNELEM strategy with the mild oscillatory problem, including details at the origin.

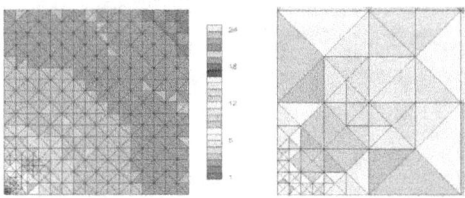

Figure 396: Example grid for the SMOOTHPRED strategy with the mild oscillatory problem, including details at the origin.

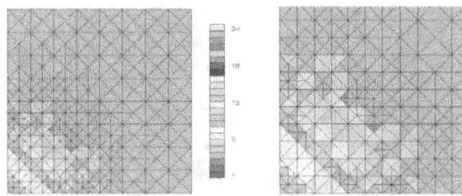

Figure 397: Example grid for the T3S strategy with the mild oscillatory problem, including details at the origin.

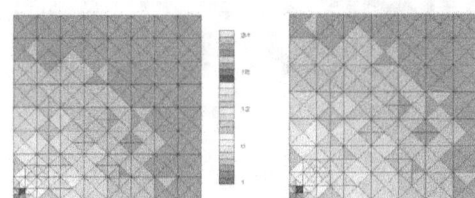

Figure 398: Example grid for the TYPEPARAM strategy with the mild oscillatory problem, including details at the origin.

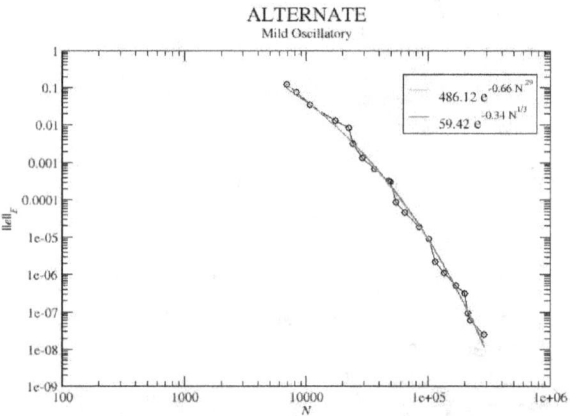

Figure 399: Log-Log plot of the convergence of the ALTERNATE strategy with the mild oscillatory problem.

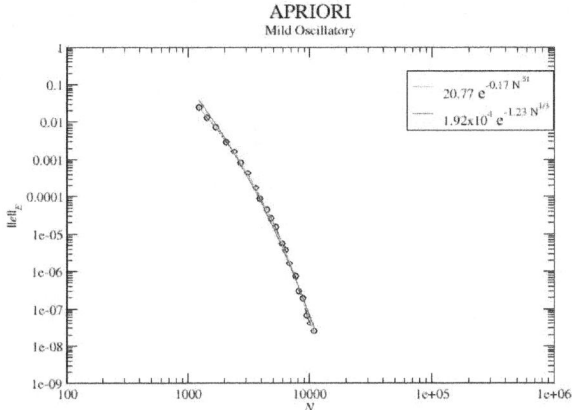

Figure 400: Log-Log plot of the convergence of the APRIORI strategy with the mild oscillatory problem.

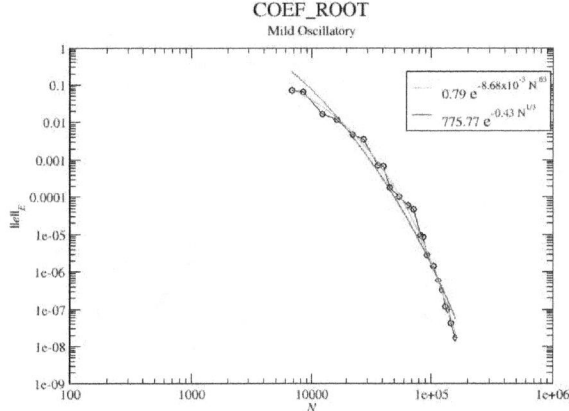

Figure 402: Log-Log plot of the convergence of the COEF-ROOT strategy with the mild oscillatory problem.

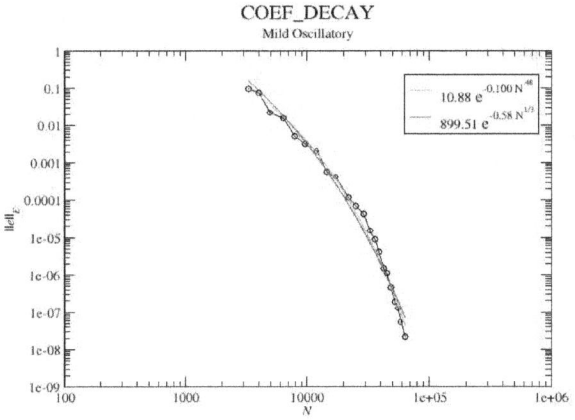

Figure 401: Log-Log plot of the convergence of the COEF-DECAY strategy with the mild oscillatory problem.

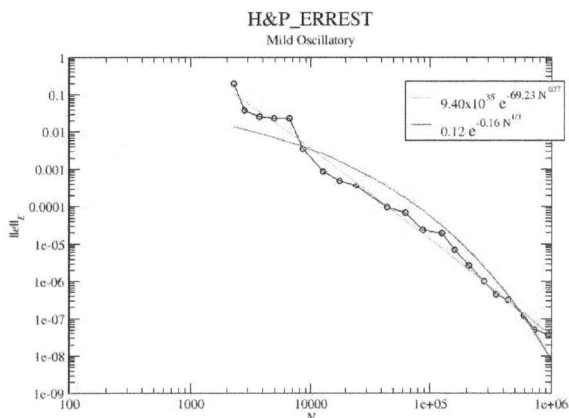

Figure 403: Log-Log plot of the convergence of the H&P-ERREST strategy with the mild oscillatory problem.

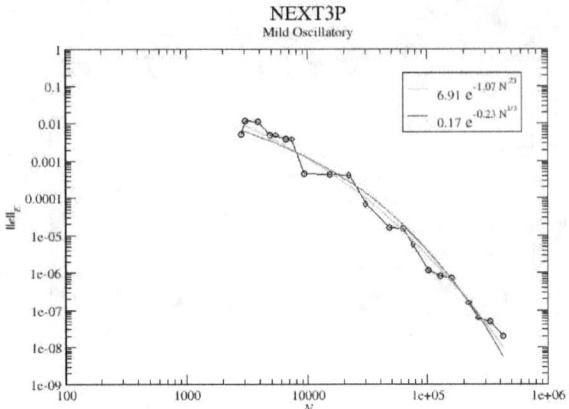
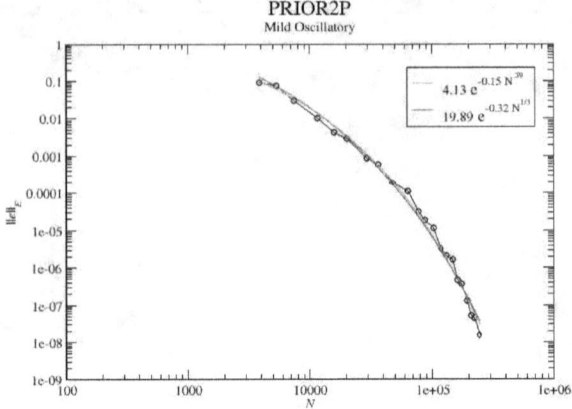

Figure 404: Log-Log plot of the convergence of the NEXT3P strategy with the mild oscillatory problem.

Figure 406: Log-Log plot of the convergence of the PRIOR2P strategy with the mild oscillatory problem.

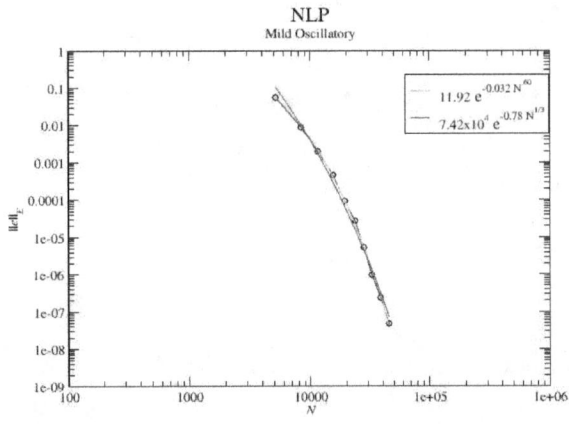
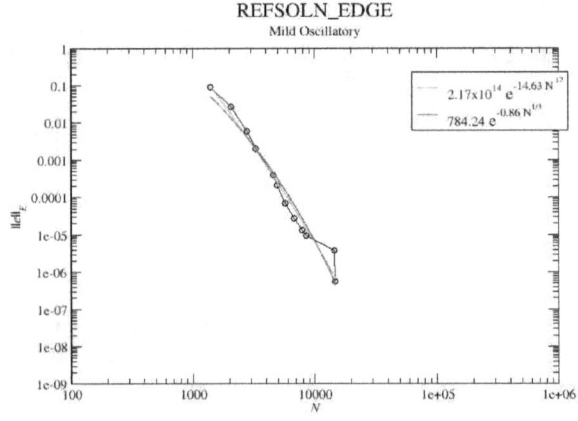

Figure 405: Log-Log plot of the convergence of the NLP strategy with the mild oscillatory problem.

Figure 407: Log-Log plot of the convergence of the REFSOLN-EDGE strategy with the mild oscillatory problem.

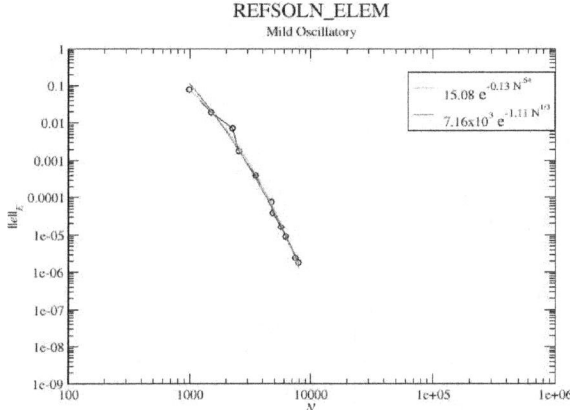
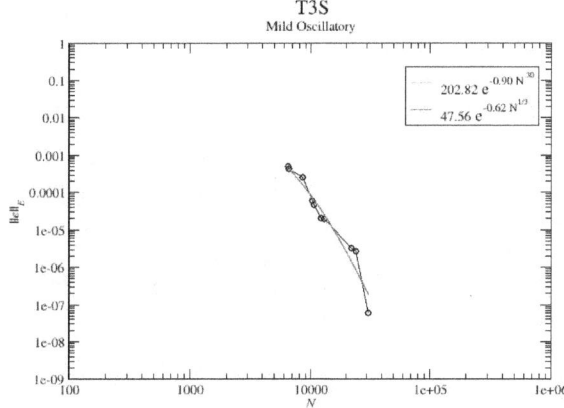

Figure 408: Log-Log plot of the convergence of the REFSOLN-ELEM strategy with the mild oscillatory problem.

Figure 410: Log-Log plot of the convergence of the T3S strategy with the mild oscillatory problem.

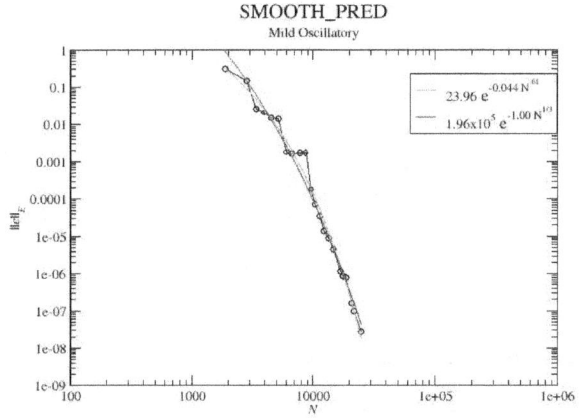
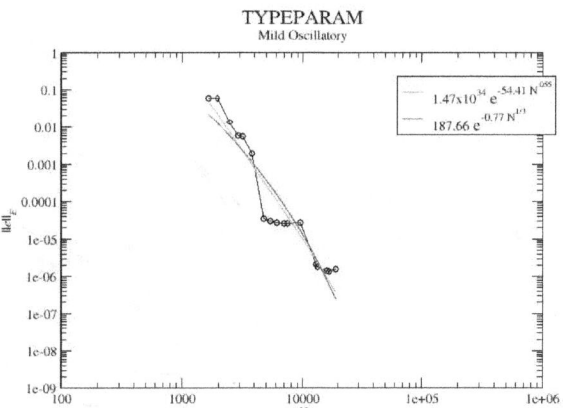

Figure 409: Log-Log plot of the convergence of the SMOOTH-PRED strategy with the mild oscillatory problem.

Figure 411: Log-Log plot of the convergence of the TYPEPARAM strategy with the mild oscillatory problem.

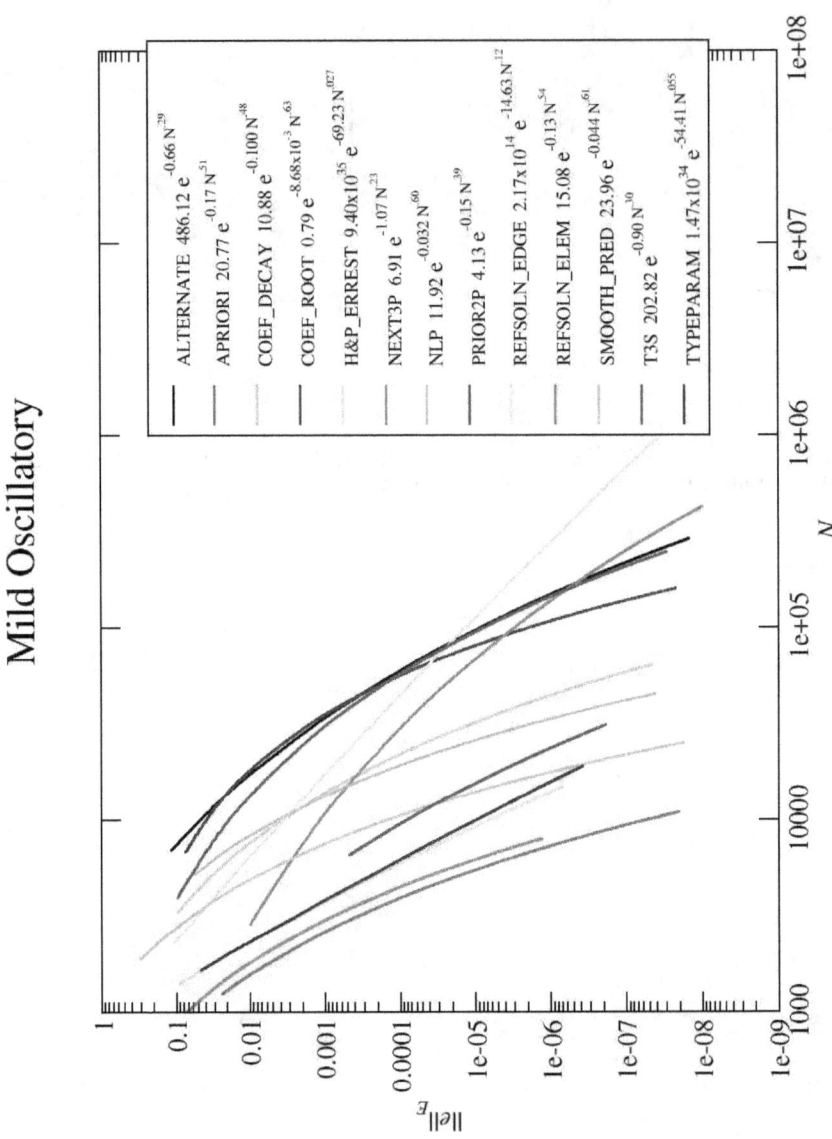

Figure 4.12: Log-Log plot of the convergence of all strategies with the mild oscillatory problem.

strategy	A	B	C
ALTERNATE	486.12	-0.66	0.29
APRIORI	20.77	-0.17	0.51
COEFDECAY	10.88	-0.100	0.48
COEFROOT	0.79	-8.68×10^{-3}	0.63
H&PERREST	9.40×10^{35}	-69.23	0.027
NEXT3P	6.91	-1.07	0.23
NLP	11.92	-0.032	0.60
PRIOR2P	4.13	-0.15	0.39
REFSOLNEDGE	2.17×10^4	-14.63	0.12
REFSOLNELEM	15.08	-0.13	0.54
SMOOTHPRED	23.96	-0.044	0.61
T3S	202.82	-0.90	0.30
TYPEPARAM	1.47×10^{34}	-54.41	0.055

Table 53: Parameters of the least squares fit for $\|e_{hp}\|_E = Ae^{BN_{dof}^C}$ for the mild oscillatory problem.

strategy	A	B
ALTERNATE	59.42	-0.34
APRIORI	1.92×10^4	-1.23
COEFDECAY	899.51	-0.58
COEFROOT	775.77	-0.43
H&PERREST	0.12	-0.16
NEXT3P	0.17	-0.23
NLP	7.42×10^4	-0.78
PRIOR2P	19.89	-0.32
REFSOLNEDGE	784.24	-0.86
REFSOLNELEM	7.16×10^3	-1.11
SMOOTHPRED	1.96×10^6	-1.00
T3S	47.56	-0.62
TYPEPARAM	187.66	-0.77

Table 55: Parameters of the least squares fit for $\|e_{hp}\|_E = Ae^{BN_{dof}^{1/3}}$ for the mild oscillatory problem.

strategy	factor
APRIORI	1.00
REFSOLNELEM	1.16
REFSOLNEDGE	1.48
TYPEPARAM	1.48
T3S	1.73
NEXT3P	1.82
SMOOTHPRED	3.12
H&PERREST	4.13
COEFDECAY	4.83
NLP	5.29
PRIOR2P	8.48
ALTERNATE	10.98
COEFROOT	11.52

Table 54: Factor by which N is larger than the best strategy for the mild oscillatory problem at low accuracy, 1.0×10^{-2}.

strategy	factor
APRIORI	1.00
REFSOLNELEM	1.12
REFSOLNEDGE	1.91
TYPEPARAM	2.15
SMOOTHPRED	2.42
T3S	3.23
NLP	4.58
COEFDECAY	6.06
COEFROOT	14.78
NEXT3P	18.81
PRIOR2P	19.79
ALTERNATE	20.32
H&PERREST	38.23

Table 56: Factor by which N is larger than the best strategy for the mild oscillatory problem at high accuracy, 1.0×10^{-6}.

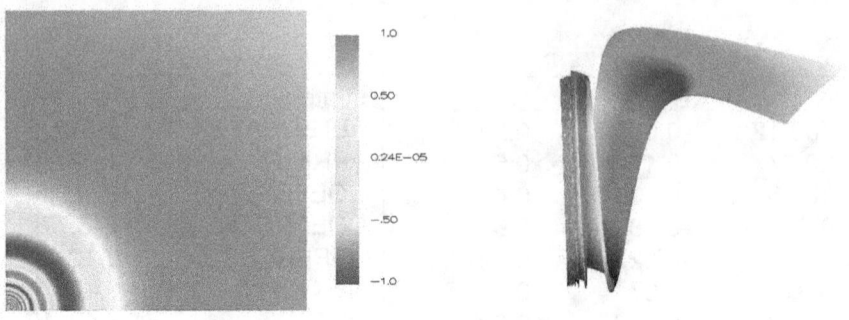

Figure 413: The solution of the strong oscillatory problem.

5.15 Oscillatory, Strong

For the strong version of the oscillatory problem (Section 5.14) we use N=50.5. τ=10^{-2} for the grid images. For APRIORI, refine by h if the element touches the origin and by p otherwise.

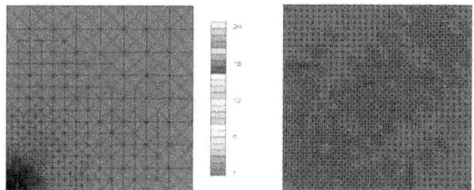

Figure 414: Example grid for the ALTERNATE strategy with the strong oscillatory problem, including details at the origin.

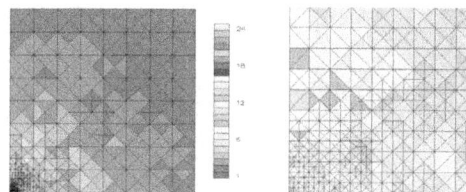

Figure 418: Example grid for the H&PERREST strategy with the strong oscillatory problem, including details at the origin.

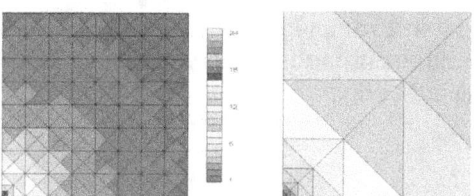

Figure 415: Example grid for the APRIORI strategy with the strong oscillatory problem, including details at the origin.

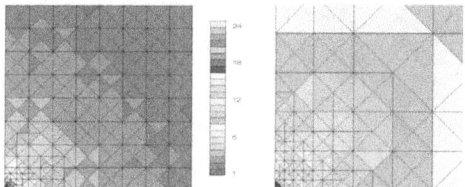

Figure 419: Example grid for the NEXT3P strategy with the strong oscillatory problem, including details at the origin.

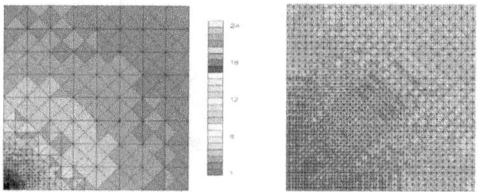

Figure 416: Example grid for the COEFDECAY strategy with the strong oscillatory problem, including details at the origin.

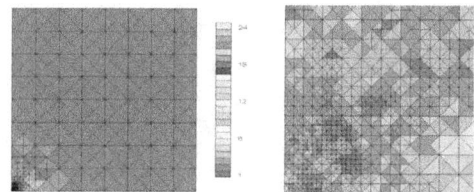

Figure 420: Example grid for the NLP strategy with the strong oscillatory problem, including details at the origin.

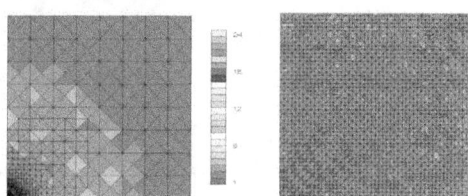

Figure 417: Example grid for the COEFROOT strategy with the strong oscillatory problem, including details at the origin.

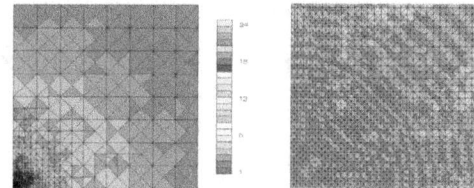

Figure 421: Example grid for the PRIOR2P strategy with the strong oscillatory problem, including details at the origin.

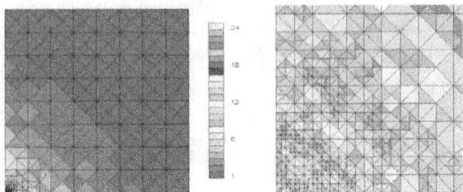

Figure 422: Example grid for the REFSOLNEDGE strategy with the strong oscillatory problem, including details at the origin.

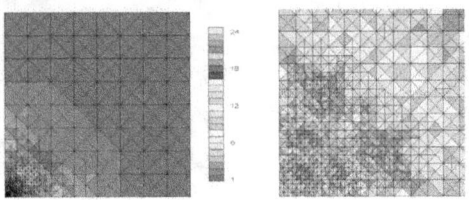

Figure 423: Example grid for the REFSOLNELEM strategy with the strong oscillatory problem, including details at the origin.

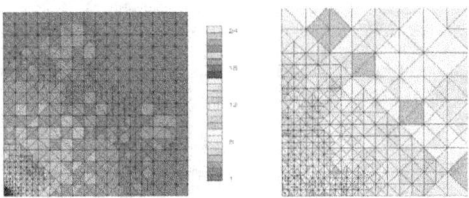

Figure 424: Example grid for the SMOOTHPRED strategy with the strong oscillatory problem, including details at the origin.

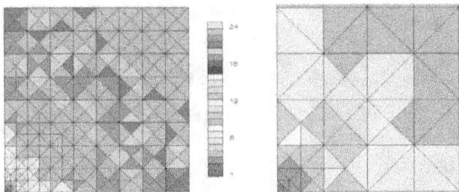

Figure 425: Example grid for the T3S strategy with the strong oscillatory problem, including details at the origin.

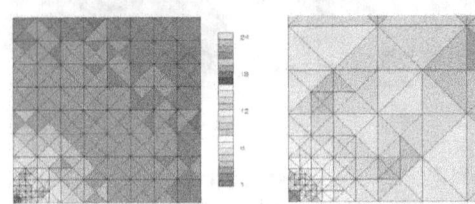

Figure 426: Example grid for the TYPEPARAM strategy with the strong oscillatory problem, including details at the origin.

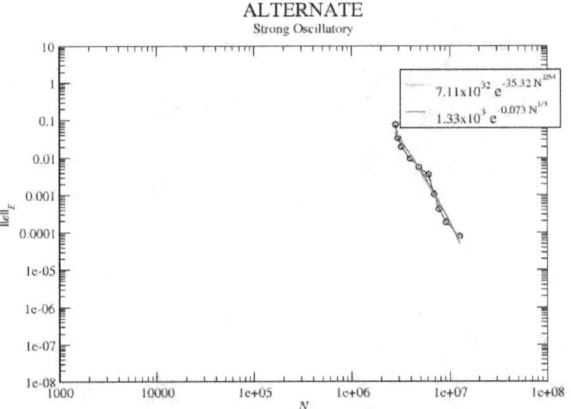

Figure 427: Log-Log plot of the convergence of the ALTERNATE strategy with the strong oscillatory problem.

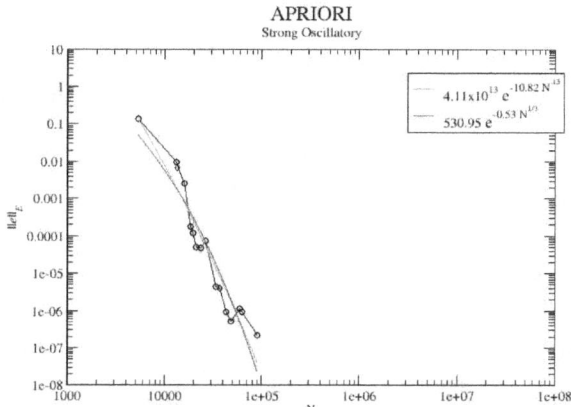

Figure 428: Log-Log plot of the convergence of the APRIORI strategy with the strong oscillatory problem.

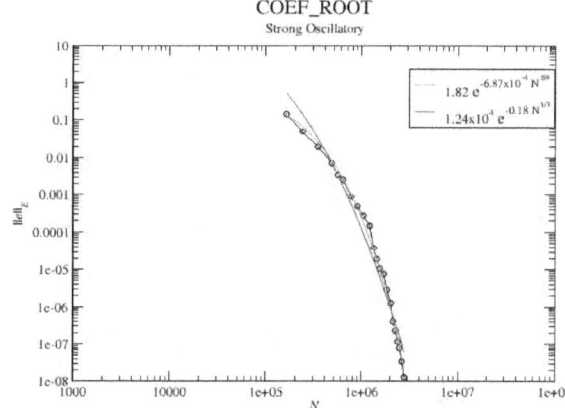

Figure 430: Log-Log plot of the convergence of the COEF-ROOT strategy with the strong oscillatory problem.

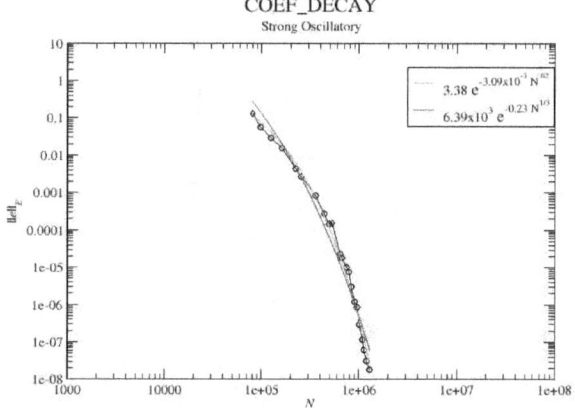

Figure 429: Log-Log plot of the convergence of the COEF-DECAY strategy with the strong oscillatory problem.

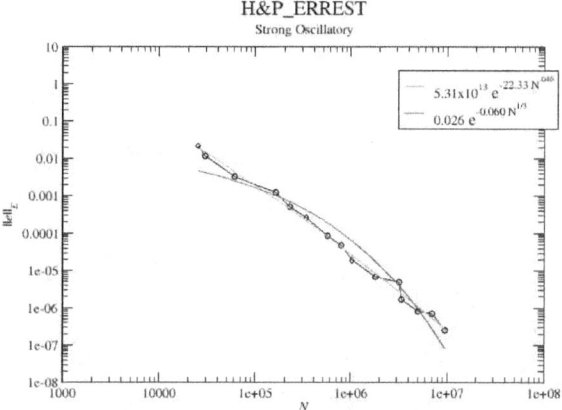

Figure 431: Log-Log plot of the convergence of the H&P-ERREST strategy with the strong oscillatory problem.

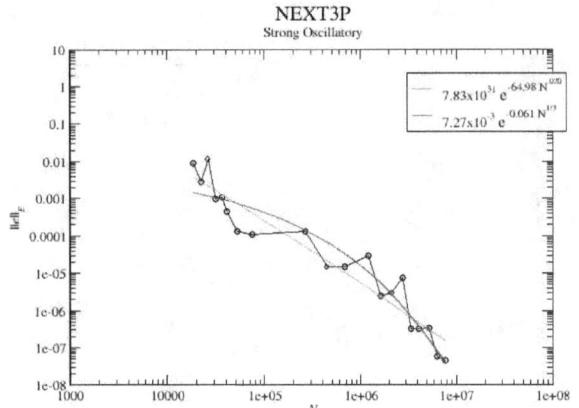

Figure 432: Log-Log plot of the convergence of the NEXT3P strategy with the strong oscillatory problem.

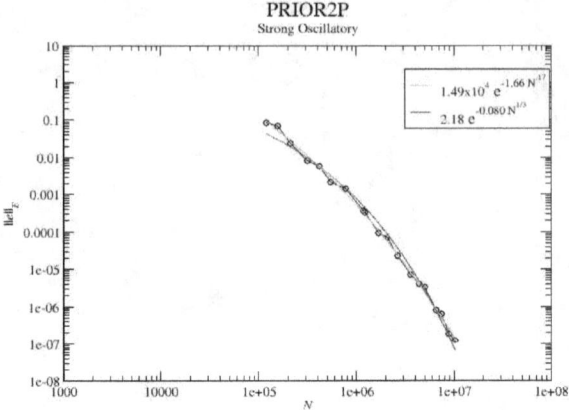

Figure 434: Log-Log plot of the convergence of the PRIOR2P strategy with the strong oscillatory problem.

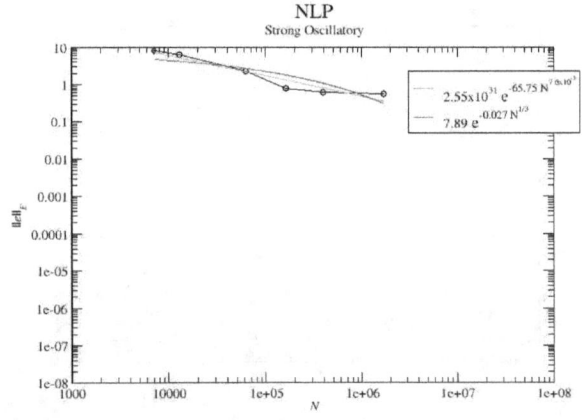

Figure 433: Log-Log plot of the convergence of the NLP strategy with the strong oscillatory problem.

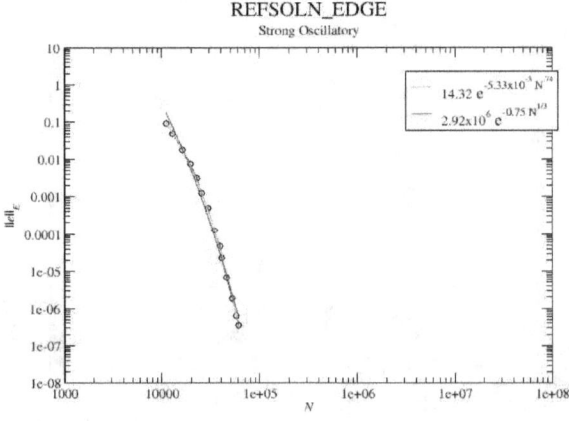

Figure 435: Log-Log plot of the convergence of the REFSOLN-EDGE strategy with the strong oscillatory problem.

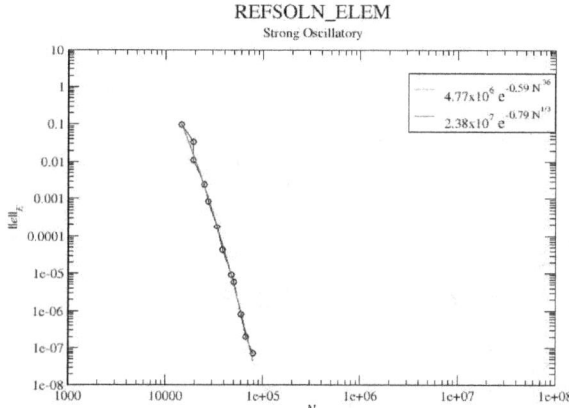

Figure 436: Log-Log plot of the convergence of the REFSOLN-ELEM strategy with the strong oscillatory problem.

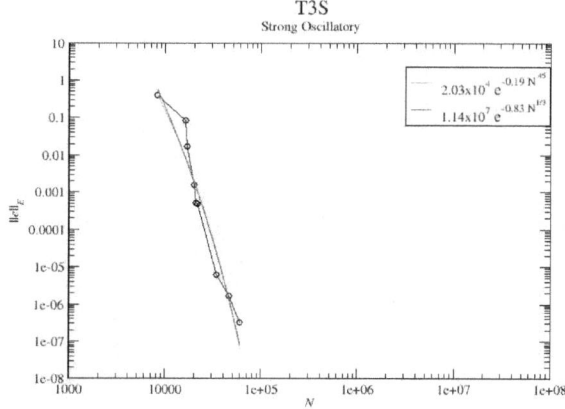

Figure 438: Log-Log plot of the convergence of the T3S strategy with the strong oscillatory problem.

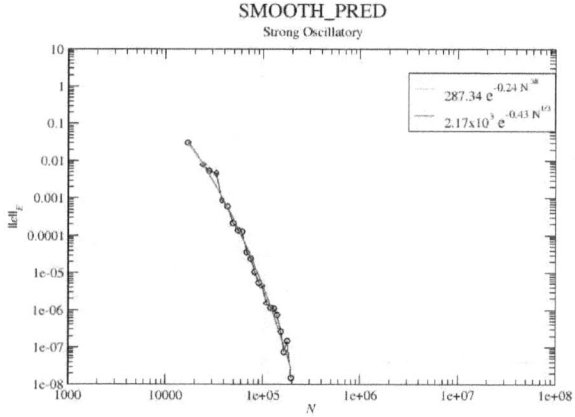

Figure 437: Log-Log plot of the convergence of the SMOOTH-PRED strategy with the strong oscillatory problem.

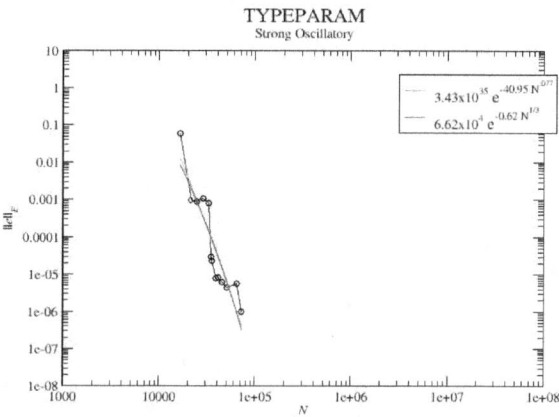

Figure 439: Log-Log plot of the convergence of the TYPEPARAM strategy with the strong oscillatory problem.

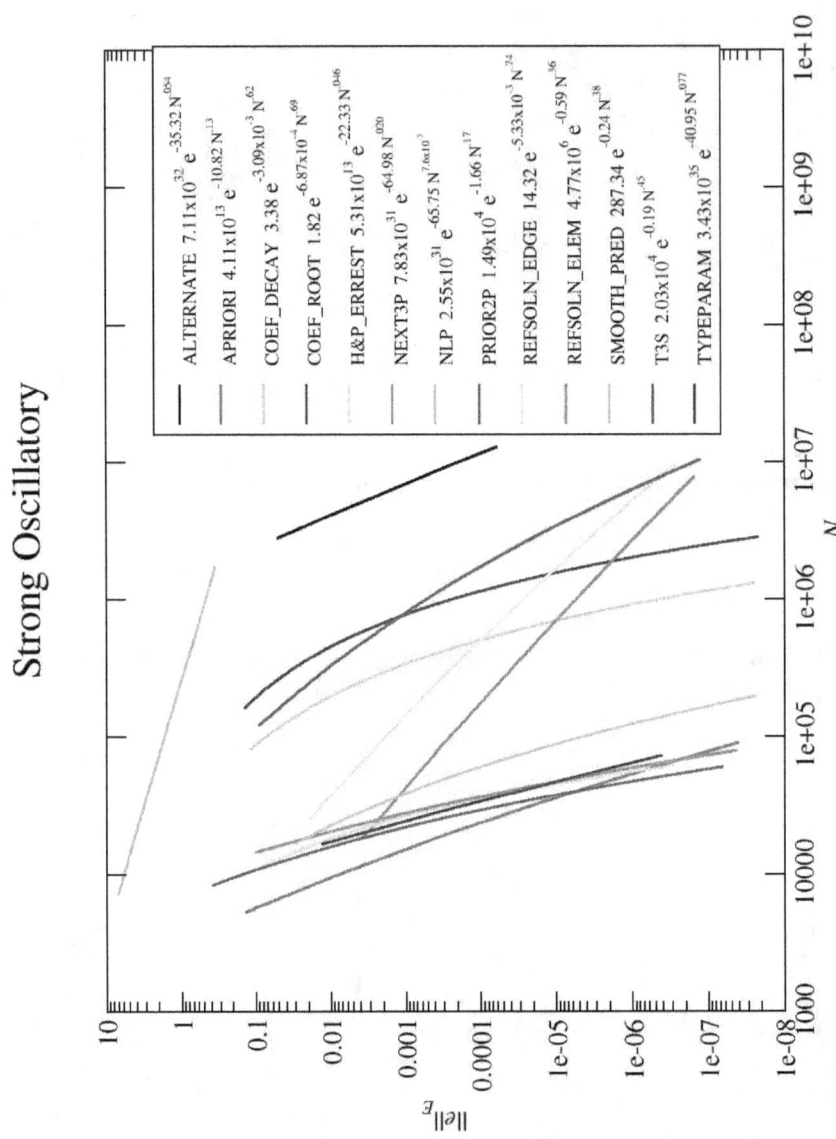

Figure 440: Log-Log plot of the convergence of all strategies with the strong oscillatory problem.

strategy	A	B	C
ALTERNATE	7.11×10^{32}	-35.32	0.054
APRIORI	4.11×10^3	-10.82	0.13
COEFDECAY	3.38	-3.09×10^{-3}	0.62
COEFROOT	1.82	-6.87×10^{-4}	0.69
H&PERREST	5.31×10^3	-22.33	0.046
NEXT3P	7.83×10^{31}	-64.98	0.020
NLP	2.55×10^{31}	-65.75	7.62×10^{-3}
PRIOR2P	1.49×10^4	-1.66	0.17
REFSOLNEDGE	14.32	-5.33×10^{-3}	0.74
REFSOLNELEM	4.77×10^6	-0.59	0.36
SMOOTHPRED	287.34	-0.24	0.38
T3S	2.03×10^4	-0.19	0.45
TYPEPARAM	3.43×10^{35}	-40.95	0.077

Table 57: Parameters of the least squares fit for $\|e_{hp}\|_E = Ae^{BN_{dof}^C}$ for the strong oscillatory problem.

strategy	A	B
ALTERNATE	1.33×10^3	-0.073
APRIORI	530.95	-0.53
COEFDECAY	6.39×10^3	-0.23
COEFROOT	1.24×10^4	-0.18
H&PERREST	0.026	-0.060
NEXT3P	7.27×10^{-3}	-0.061
NLP	7.89	-0.027
PRIOR2P	2.18	-0.080
REFSOLNEDGE	2.92×10^6	-0.75
REFSOLNELEM	2.38×10^7	-0.79
SMOOTHPRED	2.17×10^3	-0.43
T3S	1.14×10^7	-0.83
TYPEPARAM	6.62×10^4	-0.62

Table 59: Parameters of the least squares fit for $\|e_{hp}\|_E = Ae^{BN_{dof}^{1/3}}$ for the strong oscillatory problem.

strategy	factor
APRIORI	1.00
T3S	1.45
TYPEPARAM	1.60
REFSOLNEDGE	1.75
REFSOLNELEM	1.83
SMOOTHPRED	2.55
NEXT3P	2.83
H&PERREST	9.48
COEFDECAY	22.37
COEFROOT	51.26
PRIOR2P	52.76
ALTERNATE	458.20
NLP	2.78×10^6

Table 58: Factor by which N is larger than the best strategy for the strong oscillatory problem at low accuracy, 1.0×10^{-3}.

strategy	factor
T3S	1.00
APRIORI	1.14
REFSOLNEDGE	1.18
REFSOLNELEM	1.26
TYPEPARAM	1.37
SMOOTHPRED	2.62
COEFDECAY	19.39
COEFROOT	42.57
NEXT3P	57.15
H&PERREST	110.63
PRIOR2P	133.66
ALTERNATE	652.42
NLP	5.19×10^{10}

Table 60: Factor by which N is larger than the best strategy for the strong oscillatory problem at high accuracy, 1.0×10^{-6}.

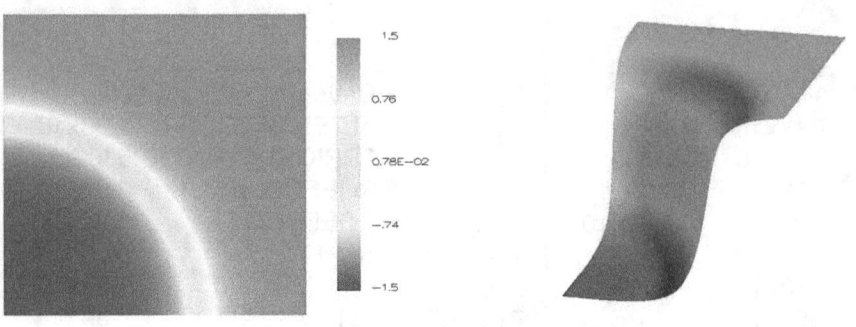

Figure 441: The solution of the mild wavefront problem.

5.16 WaveFront, Mild

The circular wavefront problem is often used as an example in adaptive grid refinement papers. It is Poisson's equation with Dirichlet boundary conditions on the unit square. The solution is

$$\tan^{-1}(\alpha(r - r_0))$$

where $r = \sqrt{(x - x_c)^2 + (y - y_c)^2}$. The location of the wavefront is defined by a circle with radius r_0 and center (x_c, y_c). α determines the steepness of the wavefront. In addition to the wavefront, the solution has a mild singularity at the center of the circle, if the center is located in the closure of the domain. For the easy form of this problem we use $\alpha=20$, $(x_c, y_c) = (-.05, -.05)$, and $r_0 = 0.7$. The center is chosen outside the domain so that only the wavefront is a factor in the adaptivity, not the singularity. $\tau = 10^{-4}$ for the grid images. For the APRIORI strategy, refine by h if the element touches the circle that defines the location of the wavefront and has degree at least 3 (chosen arbitrarily), and by p otherwise.

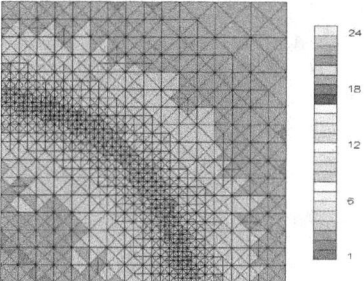

Figure 442: Example grid for the ALTERNATE strategy with the mild wavefront problem.

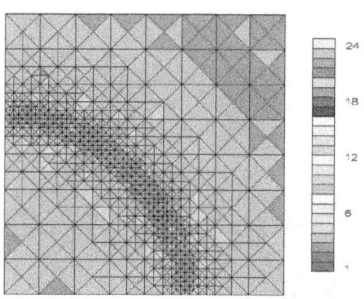

Figure 445: Example grid for the COEFROOT strategy with the mild wavefront problem.

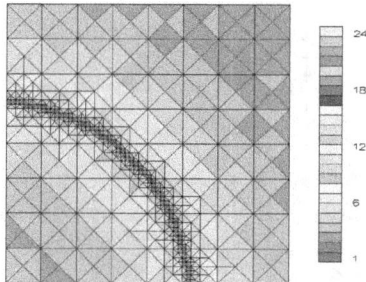

Figure 443: Example grid for the APRIORI strategy with the mild wavefront problem.

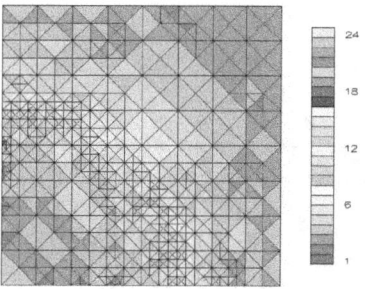

Figure 446: Example grid for the H&PERREST strategy with the mild wavefront problem.

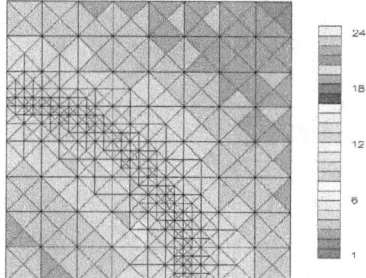

Figure 444: Example grid for the COEFDECAY strategy with the mild wavefront problem.

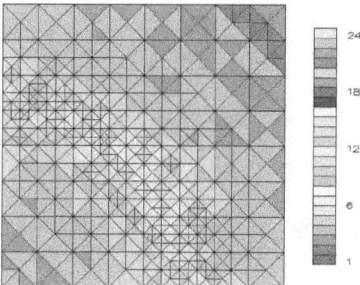

Figure 447: Example grid for the NEXT3P strategy with the mild wavefront problem.

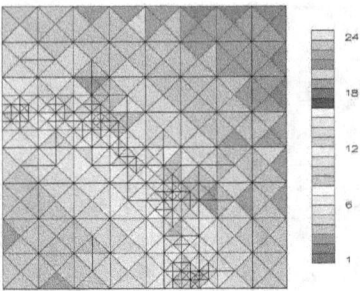

Figure 448: Example grid for the NLP strategy with the mild wavefront problem.

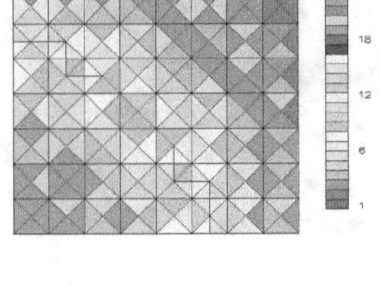

Figure 451: Example grid for the REFSOLNELEM strategy with the mild wavefront problem.

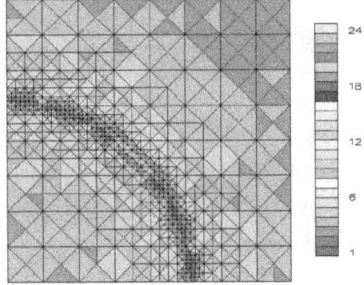

Figure 449: Example grid for the PRIOR2P strategy with the mild wavefront problem.

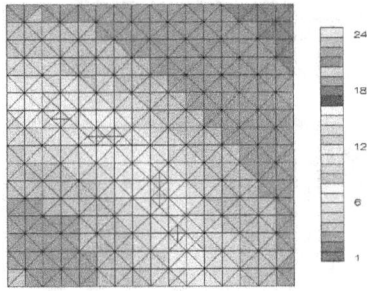

Figure 452: Example grid for the SMOOTHPRED strategy with the mild wavefront problem.

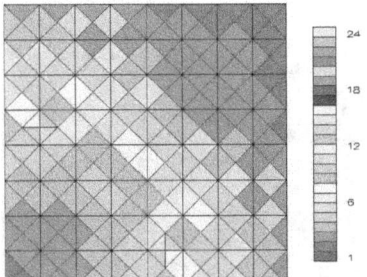

Figure 450: Example grid for the REFSOLNEDGE strategy with the mild wavefront problem.

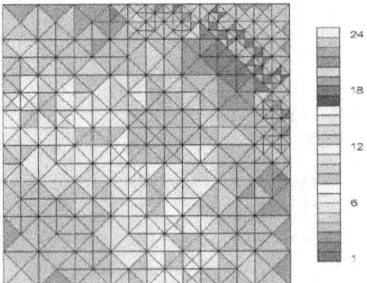

Figure 453: Example grid for the T3S strategy with the mild wavefront problem.

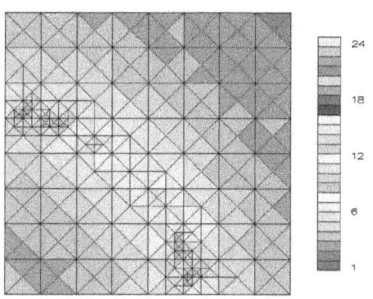

Figure 454: Example grid for the TYPEPARAM strategy with the mild wave front problem.

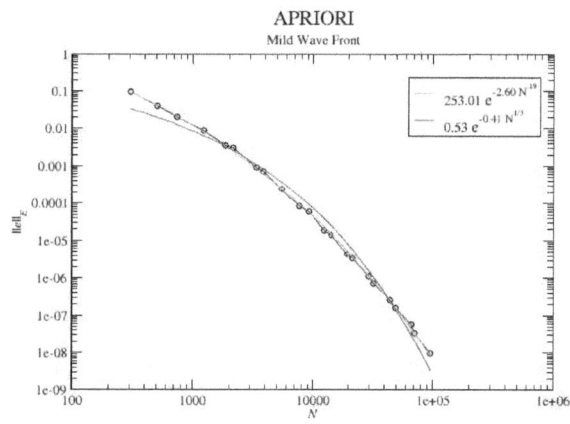

Figure 456: Log-Log plot of the convergence of the APRIORI strategy with the mild wave front problem.

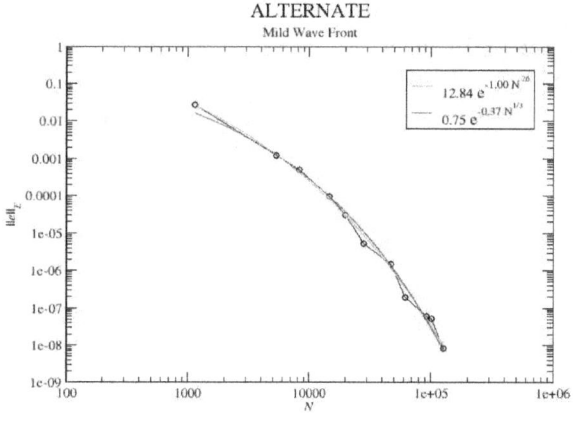

Figure 455: Log-Log plot of the convergence of the ALTERNATE strategy with the mild wave front problem.

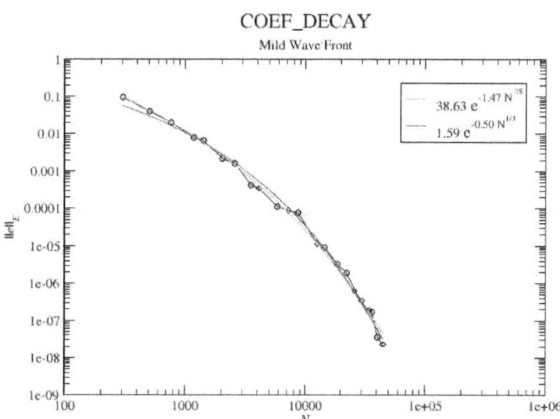

Figure 457: Log-Log plot of the convergence of the COEF-DECAY strategy with the mild wave front problem.

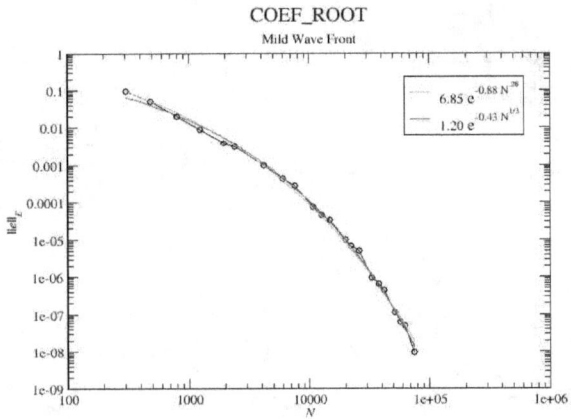
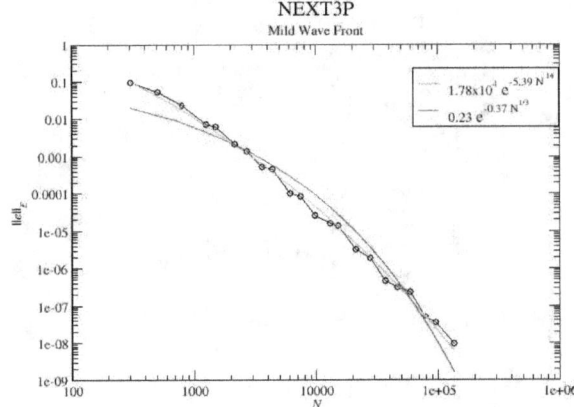

Figure 458: Log-Logplot of the convergence of the COEF-ROOT strategy with the mild wave front problem.

Figure 460: Log-Logplot of the convergence of the NEXT3P strategy with the mild wave front problem.

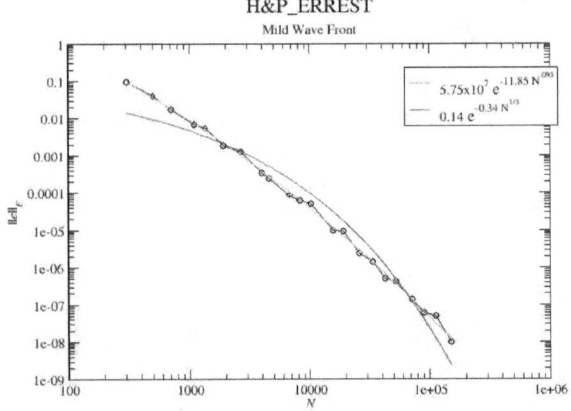
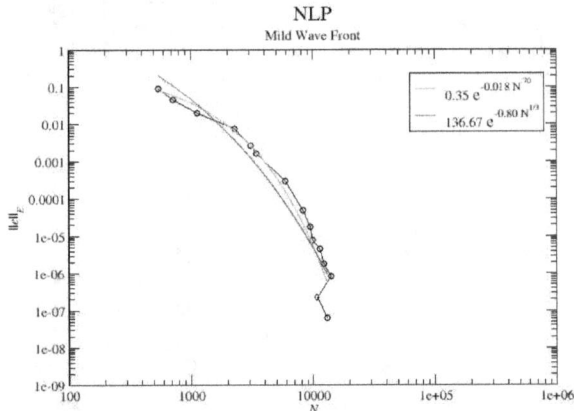

Figure 459: Log-Logplot of the convergence of the H&P-ERREST strategy with the mild wave front problem.

Figure 461: Log-Logplot of the convergence of the NLP strategy with the mild wave front problem.

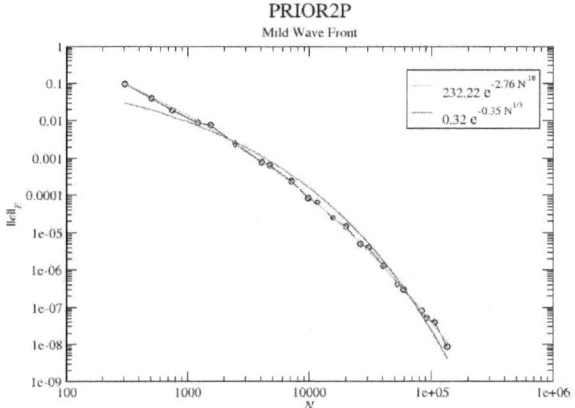

Figure 462: Log-Log plot of the convergence of the PRIOR2P strategy with the mild wave front problem.

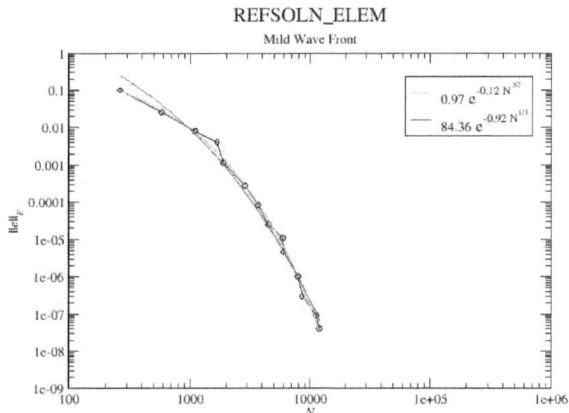

Figure 464: Log-Logplot of the convergence of the REFSOLN-ELEM strategy with the mild wave front problem.

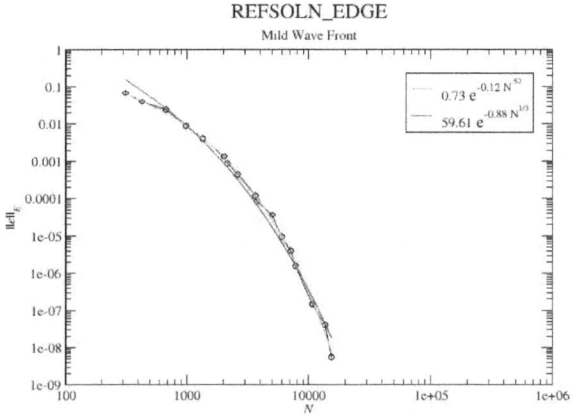

Figure 463: Log-Logplot of the convergence of the REFSOLN-EDGE strategy with the mild wave front problem.

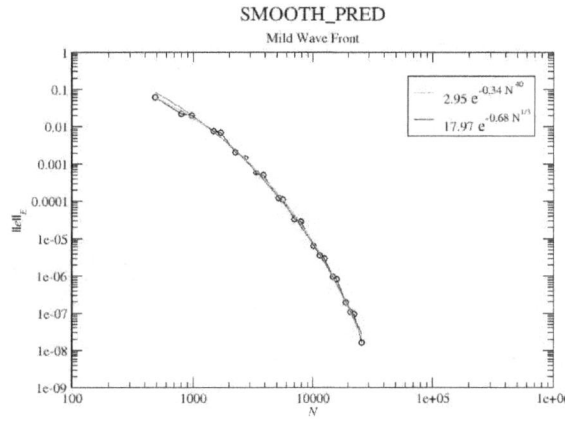

Figure 465: Log-Logplot of the convergence of the SMOOTH-PRED strategy with the mild wave front problem.

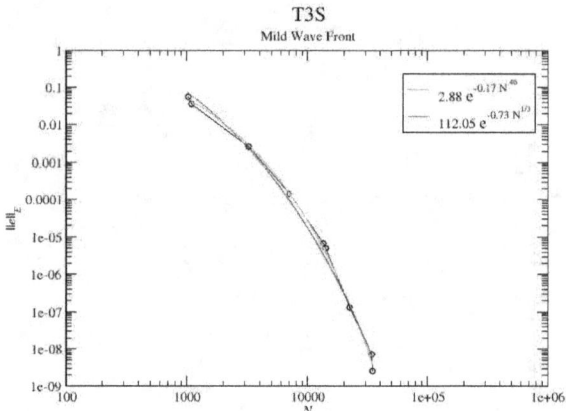

Figure 466: Log-Log plot of the convergence of the T3S strategy with the mild wave front problem.

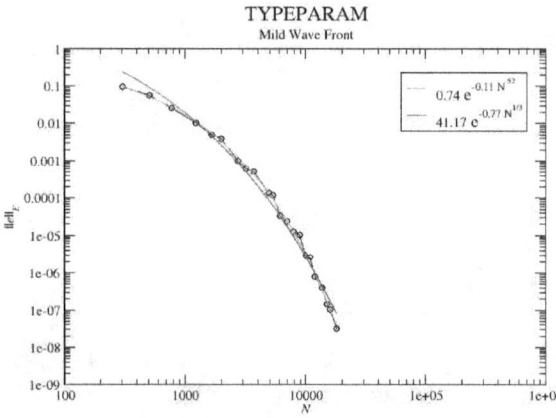

Figure 467: Log-Log plot of the convergence of the TYPEPARAM strategy with the mild wave front problem.

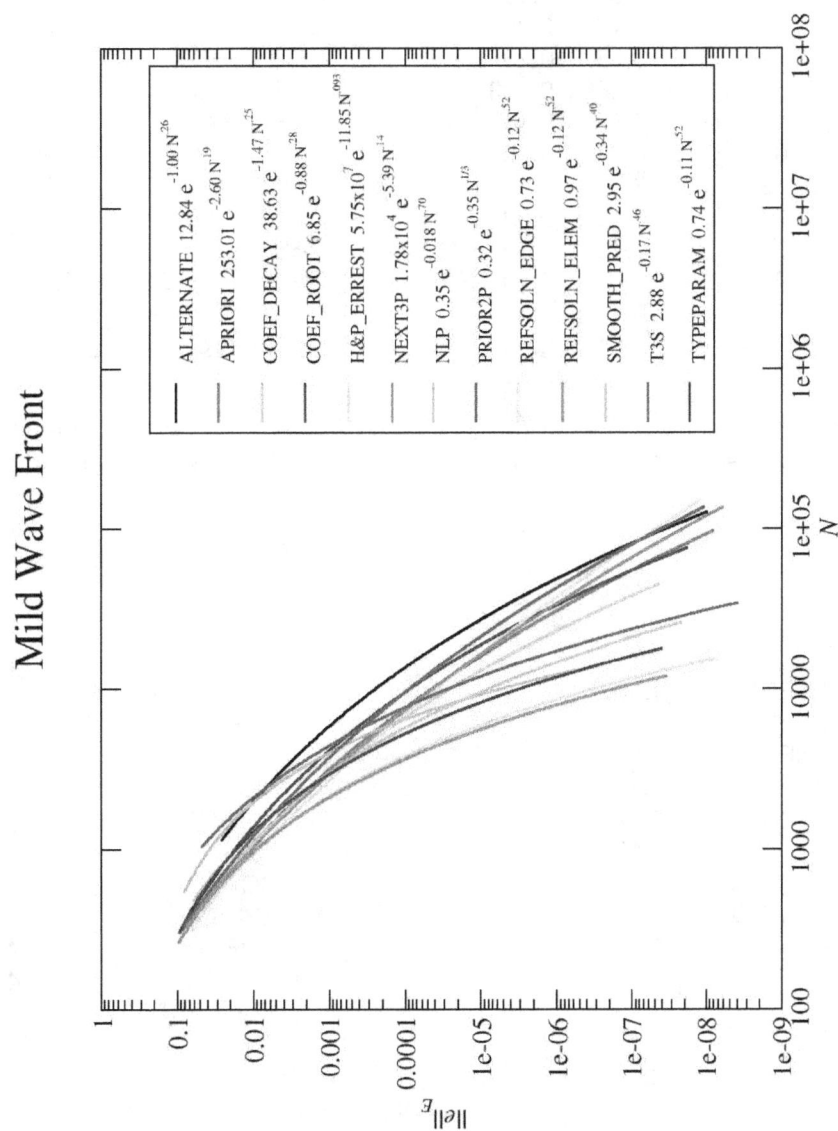

Figure 468: Log-Log plot of the convergence of all strategies with the mild wave front problem.

strategy	A	B	C
ALTERNATE	12.84	-1.00	0.26
APRIORI	253.01	-2.60	0.19
COEFDECAY	38.63	-1.47	0.25
COEFROOT	6.85	-0.88	0.28
H&PERREST	5.75×10^7	-11.85	0.093
NEXT3P	1.78×10^4	-5.39	0.14
NLP	0.35	-0.018	0.70
PRIOR2P	232.22	-2.76	0.18
REFSOLNEDGE	0.73	-0.12	0.52
REFSOLNELEM	0.97	-0.12	0.52
SMOOTHPRED	2.95	-0.34	0.40
T3S	2.88	-0.17	0.46
TYPEPARAM	0.74	-0.11	0.52

Table 61: Parameters of the least squares fit for $\|e_{hp}\|_E = Ae^{BN_{dof}^C}$ for the mild wavefront problem.

strategy	A	B
ALTERNATE	0.75	-0.37
APRIORI	0.53	-0.41
COEFDECAY	1.59	-0.50
COEFROOT	1.20	-0.43
H&PERREST	0.14	-0.34
NEXT3P	0.23	-0.37
NLP	136.67	-0.80
PRIOR2P	0.32	-0.35
REFSOLNEDGE	59.61	-0.88
REFSOLNELEM	84.36	-0.92
SMOOTHPRED	17.97	-0.68
T3S	112.05	-0.73
TYPEPARAM	41.17	-0.77

Table 63: Parameters of the least squares fit for $\|e_{hp}\|_E = Ae^{BN_{dof}^{1/3}}$ for the mild wavefront problem.

strategy	factor
REFSOLNEDGE	1.00
H&PERREST	1.01
REFSOLNELEM	1.03
NEXT3P	1.10
COEFDECAY	1.17
APRIORI	1.17
PRIOR2P	1.28
SMOOTHPRED	1.36
TYPEPARAM	1.37
COEFROOT	1.48
NLP	2.09
ALTERNATE	2.13
T3S	2.24

Table 62: Factor by which N is larger than the best strategy for the mild wavefront problem at low accuracy, 1.0×10^{-2}.

strategy	factor
REFSOLNELEM	1.00
REFSOLNEDGE	1.08
TYPEPARAM	1.49
NLP	1.55
SMOOTHPRED	1.85
T3S	2.18
COEFDECAY	2.91
APRIORI	3.88
COEFROOT	4.22
NEXT3P	4.32
H&PERREST	4.73
PRIOR2P	5.41
ALTERNATE	6.15

Table 64: Factor by which N is larger than the best strategy for the mild wavefront problem at high accuracy, 1.0×10^{-6}.

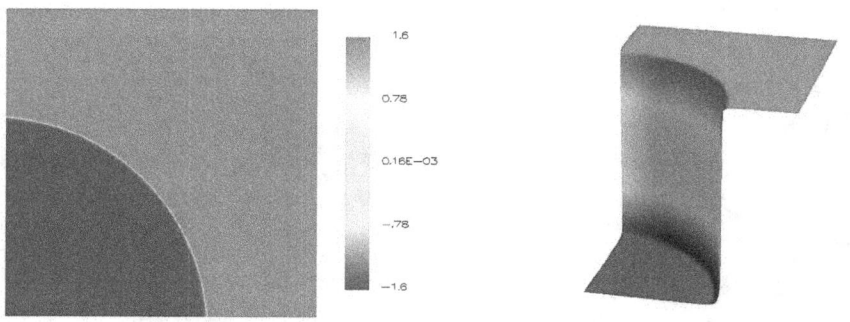

Figure 469: The solution of the steep wavefront problem.

5.17 WaveFront, Steep

In the hard version of the wavefront problem (Section 5.16) the location of the wavefront is the same, but it is much steeper. The parameters are $\alpha=1000$, $(x_c, y_c) = (-.05, -.05)$, and $r_0 = 0.7$. $\tau = 10^{-1}$ for the grid images. For the APRIORI strategy, refine by h if the element touches the circle that defines the location of the wavefront and has degree at least 3 (chosen arbitrarily), and by p otherwise.

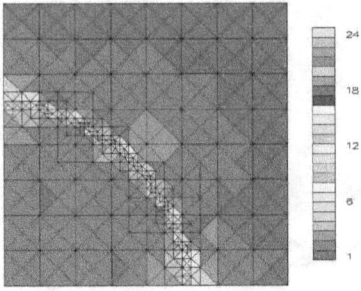
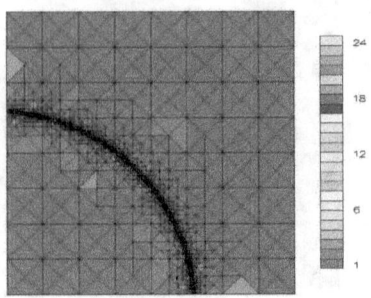

Figure 470: Example grid for the ALTERNATE strategy with the steep wavefront problem.

Figure 473: Example grid for the COEFROOT strategy with the steep wavefront problem.

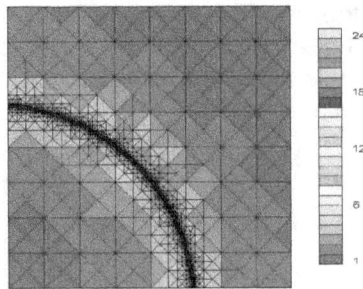
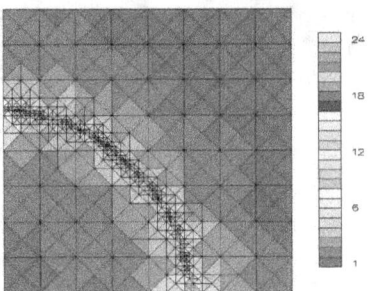

Figure 471: Example grid for the APRIORI strategy with the steep wavefront problem.

Figure 474: Example grid for the H&PERREST strategy with the steep wavefront problem.

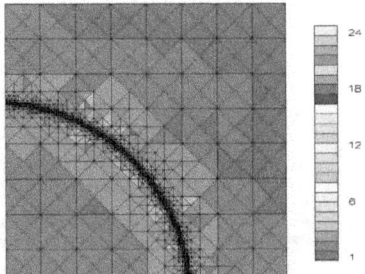
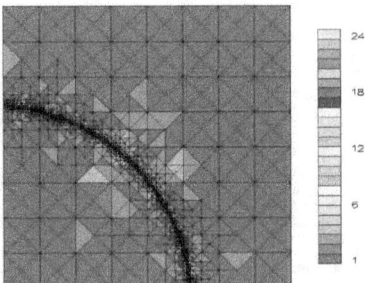

Figure 472: Example grid for the COEFDECAY strategy with the steep wavefront problem.

Figure 475: Example grid for the NEXT3P strategy with the steep wavefront problem.

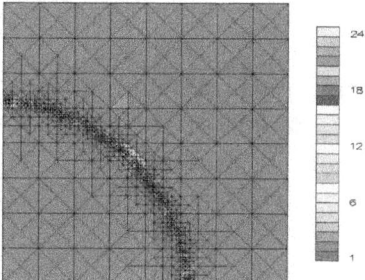

Figure 476: Example grid for the NLP strategy with the steep wavefront problem.

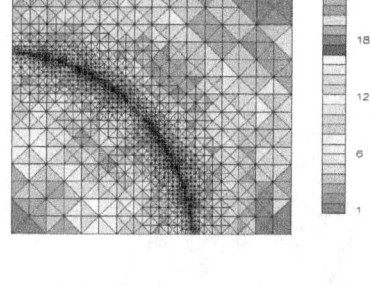

Figure 479: Example grid for the REFSOLNELEM strategy with the steep wavefront problem.

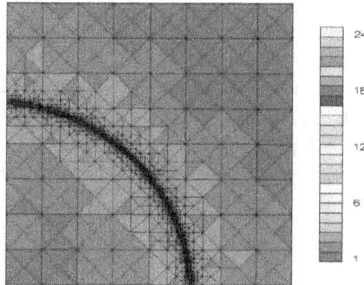

Figure 477: Example grid for the PRIOR2P strategy with the steep wavefront problem.

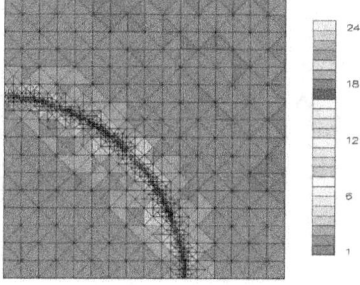

Figure 480: Example grid for the SMOOTHPRED strategy with the steep wavefront problem.

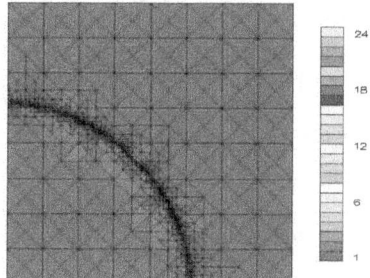

Figure 478: Example grid for the REFSOLNEDGE strategy with the steep wavefront problem.

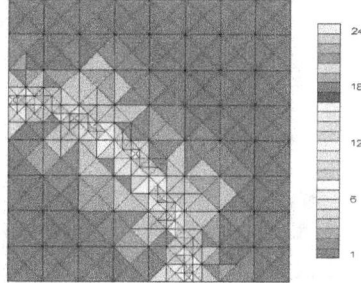

Figure 481: Example grid for the T3S strategy with the steep wavefront problem.

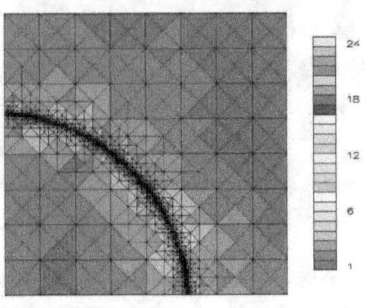

Figure 482: Example grid for the TYPEPARAM strategy with the steep wave front problem.

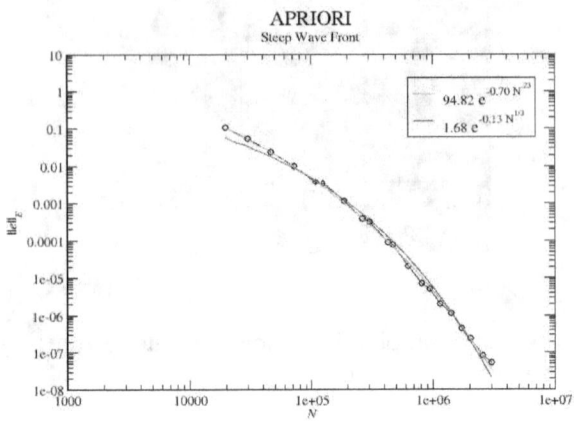

Figure 484: Log-Log plot of the convergence of the APRIORI strategy with the steep wave front problem.

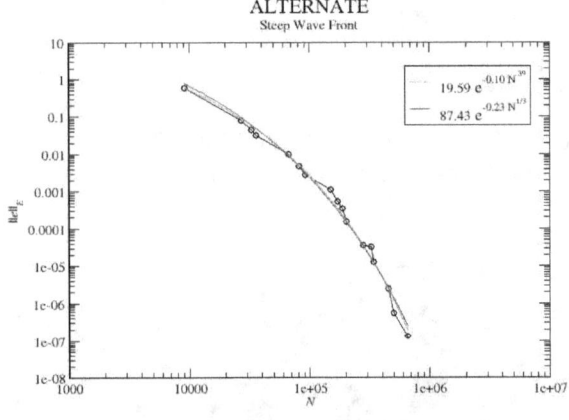

Figure 483: Log-Log plot of the convergence of the ALTERNATE strategy with the steep wave front problem.

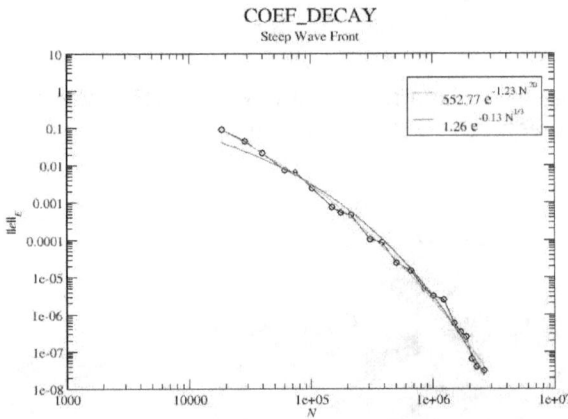

Figure 485: Log-Log plot of the convergence of the COEF-DECAY strategy with the steep wave front problem.

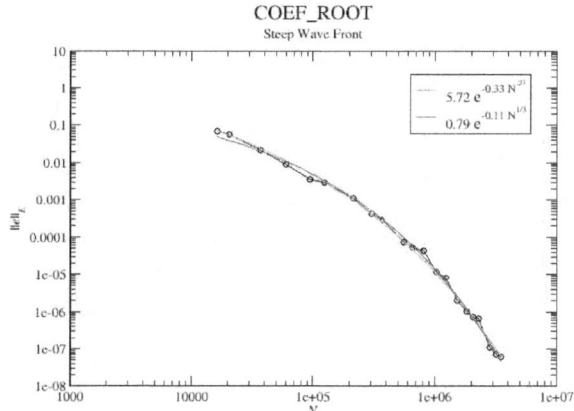

Figure 486: Log-Logplot of the convergence of the COEF-ROOT strategy with the steep wave front problem.

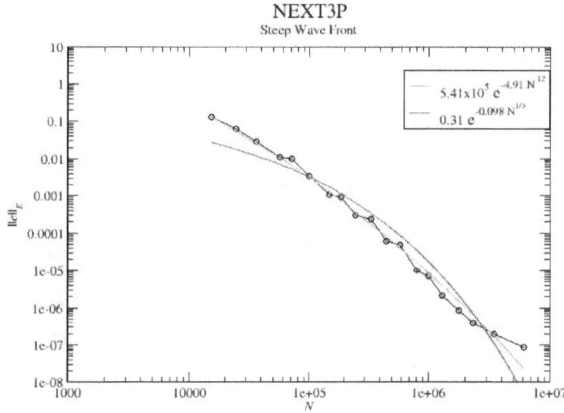

Figure 488: Log-Logplot of the convergence of the NEXT3P strategy with the steep wave front problem.

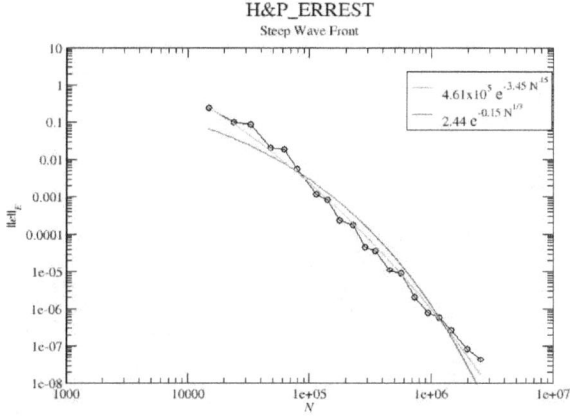

Figure 487: Log-Logplot of the convergence of the H&P-ERREST strategy with the steep wave front problem.

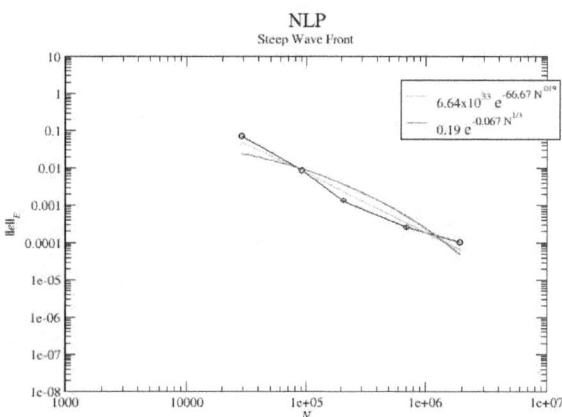

Figure 489: Log-Logplot of the convergence of the NLP strategy with the steep wave front problem.

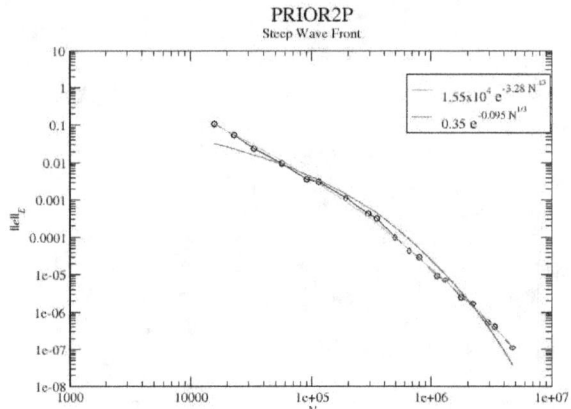

Figure 490: Log-Log plot of the convergence of the PRIOR2P strategy with the steep wave front problem.

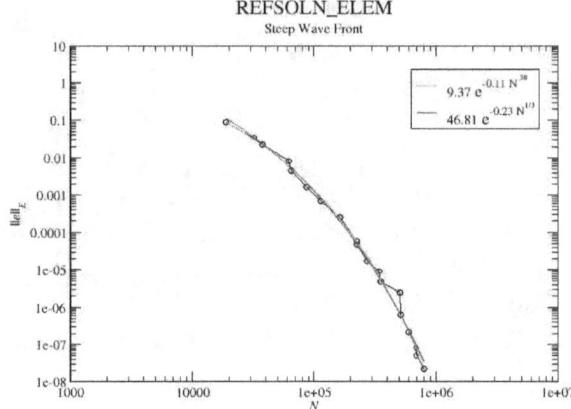

Figure 492: Log-Log plot of the convergence of the REFSOLN_ELEM strategy with the steep wave front problem.

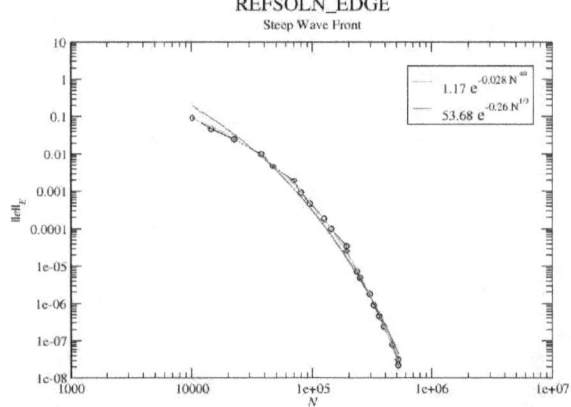

Figure 491: Log-Log plot of the convergence of the REFSOLN_EDGE strategy with the steep wave front problem.

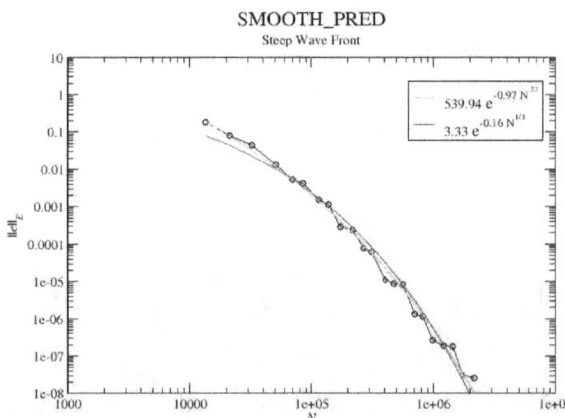

Figure 493: Log-Log plot of the convergence of the SMOOTH_PRED strategy with the steep wave front problem.

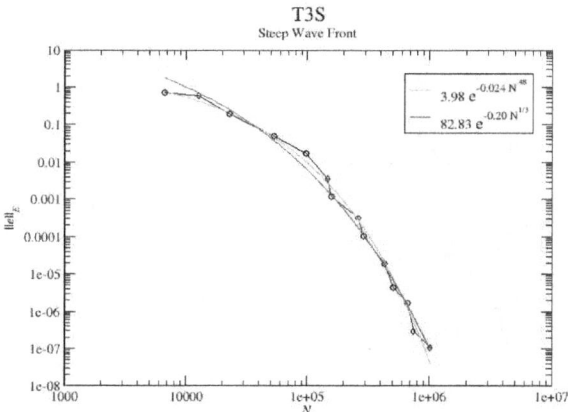

Figure 494: Log-Log plot of the convergence of the T3S strategy with the steep wave front problem.

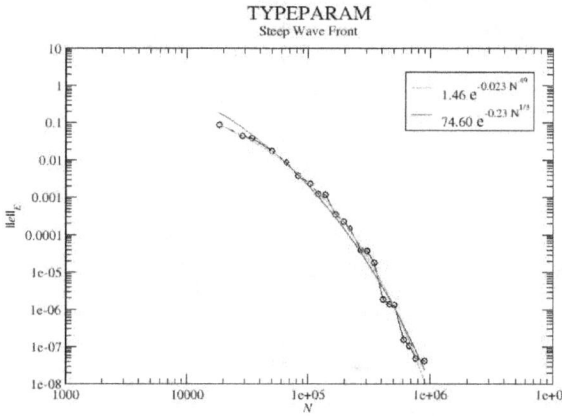

Figure 495: Log-Log plot of the convergence of the TYPEPARAM strategy with the steep wave front problem.

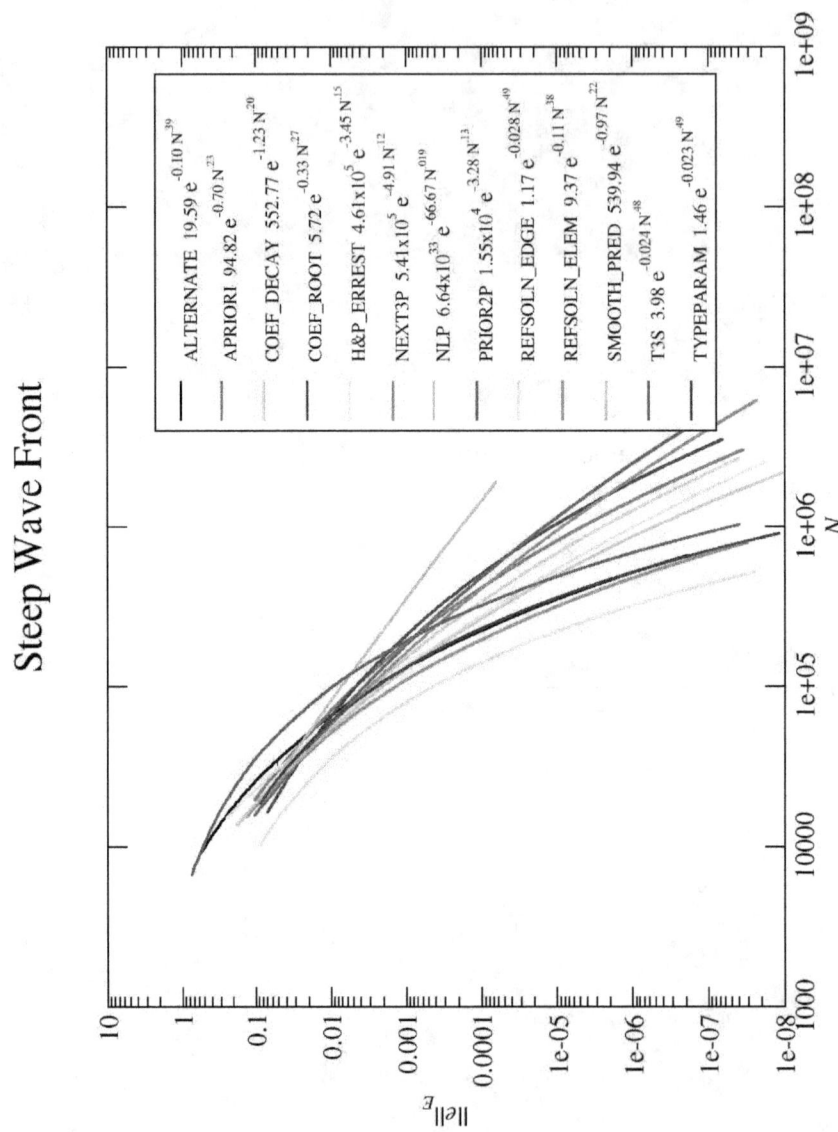

Figure 496: Log-Log plot of the convergence of all strategies with the steep wave front problem.

strategy	A	B	C
ALTERNATE	19.59	-0.10	0.39
APRIORI	94.82	-0.70	0.23
COEFDECAY	552.77	-1.23	0.20
COEFROOT	5.72	-0.33	0.27
H&PERREST	4.61×10^6	-3.45	0.15
NEXT3P	5.41×10^6	-4.91	0.12
NLP	6.64×10^{33}	-66.67	0.019
PRIOR2P	1.55×10^4	-3.28	0.13
REFSOLNEDGE	1.17	-0.028	0.49
REFSOLNELEM	9.37	-0.11	0.38
SMOOTHPRED	539.94	-0.97	0.22
T3S	3.98	-0.024	0.48
TYPEPARAM	1.46	-0.023	0.49

Table 65: Parameters of the least squares fit for $\|e_{hp}\|_E = Ae^{BN_{dof}^C}$ for the steep wavefront problem.

strategy	A	B
ALTERNATE	87.43	-0.23
APRIORI	1.68	-0.13
COEFDECAY	1.26	-0.13
COEFROOT	0.79	-0.11
H&PERREST	2.44	-0.15
NEXT3P	0.31	-0.098
NLP	0.19	-0.067
PRIOR2P	0.35	-0.095
REFSOLNEDGE	53.68	-0.26
REFSOLNELEM	46.81	-0.23
SMOOTHPRED	3.33	-0.16
T3S	82.83	-0.20
TYPEPARAM	74.60	-0.23

Table 67: Parameters of the least squares fit for $\|e_{hp}\|_E = Ae^{BN_{dof}^{1/3}}$ for the steep wavefront problem.

strategy	factor
REFSOLNEDGE	1.00
REFSOLNELEM	1.45
SMOOTHPRED	1.53
NEXT3P	1.63
COEFDECAY	1.64
H&PERREST	1.65
PRIOR2P	1.68
TYPEPARAM	1.70
COEFROOT	1.84
ALTERNATE	1.85
APRIORI	2.00
NLP	2.30
T3S	2.76

Table 66: Factor by which N is larger than the best strategy for the steep wavefront problem at low accuracy, 1.0×10^{-2}.

strategy	factor
REFSOLNEDGE	1.00
REFSOLNELEM	1.48
ALTERNATE	1.61
TYPEPARAM	1.61
T3S	2.14
SMOOTHPRED	2.69
H&PERREST	3.09
COEFDECAY	3.93
APRIORI	4.55
COEFROOT	5.94
NEXT3P	6.31
PRIOR2P	7.74
NLP	70.96

Table 68: Factor by which N is larger than the best strategy for the steep wavefront problem at high accuracy, 1.0×10^{-6}.

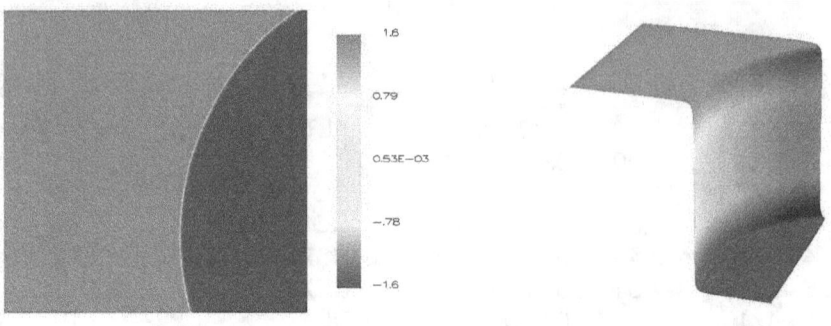

Figure 497: The solution of the asymmetric wavefront problem.

5.18 WaveFront, Asymmetric

The asymmetric wavefront is similar to the steep wavefront except the wavefront is not symmetric within the domain. The parameters are $\alpha=1000$, $(x_c,y_c)=(1.5,.25)$, and $r_0=.92$. $\tau=10^{-1}$ for the grid images. For the APRIORI strategy, refine by h if the element touches the circle that defines the location of the wave front and has degree at least 3 (chosen arbitrarily), and by p otherwise.

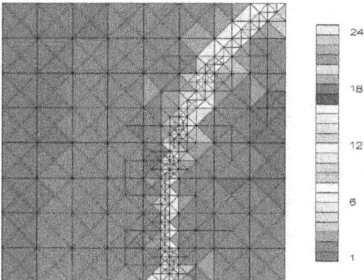

Figure 498: Example grid for the ALTERNATE strategy with the asymmetric wavefront problem.

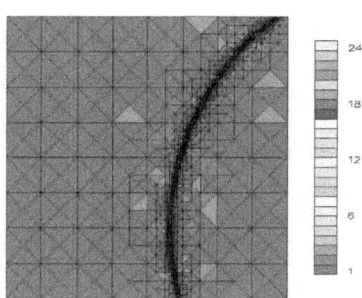

Figure 501: Example grid for the COEFROOT strategy with the asymmetric wavefront problem.

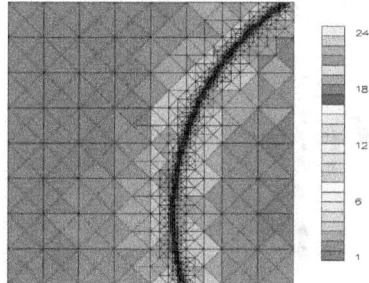

Figure 499: Example grid for the APRIORI strategy with the asymmetric wavefront problem.

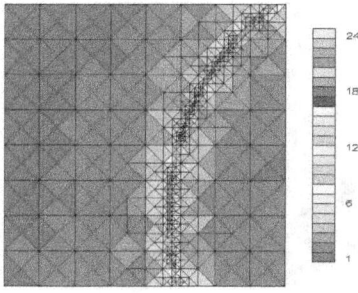

Figure 502: Example grid for the H&PERREST strategy with the asymmetric wavefront problem.

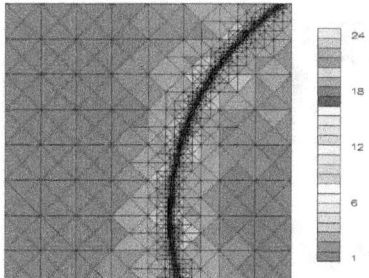

Figure 500: Example grid for the COEFDECAY strategy with the asymmetric wavefront problem.

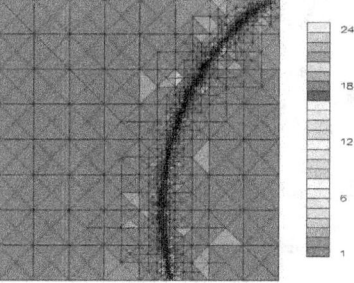

Figure 503: Example grid for the NEXT3P strategy with the asymmetric wavefront problem.

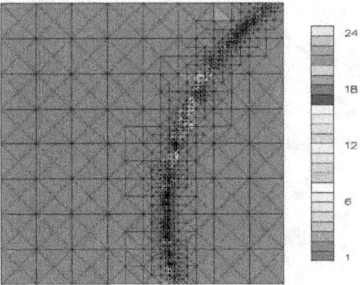

Figure 504: Example grid for the NLP strategy with the asymmetric wavefront problem.

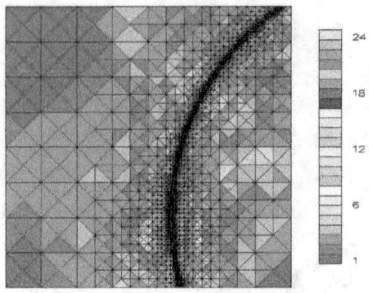

Figure 507: Example grid for the REFSOLNELEM strategy with the asymmetric wavefront problem.

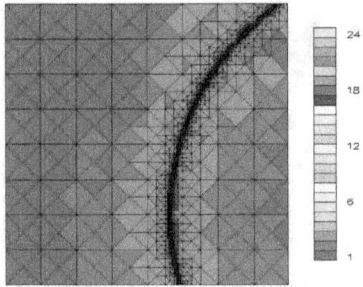

Figure 505: Example grid for the PRIOR2P strategy with the asymmetric wavefront problem.

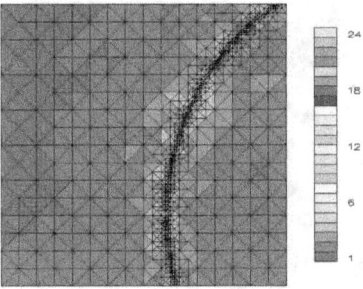

Figure 508: Example grid for the SMOOTHPRED strategy with the asymmetric wavefront problem.

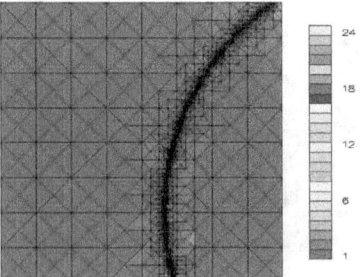

Figure 506: Example grid for the REFSOLNEDGE strategy with the asymmetric wavefront problem.

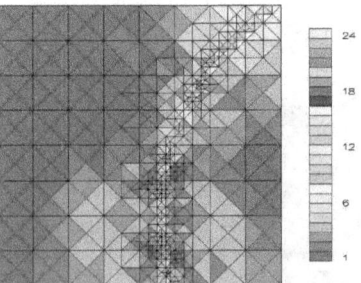

Figure 509: Example grid for the T3S strategy with the asymmetric wavefront problem.

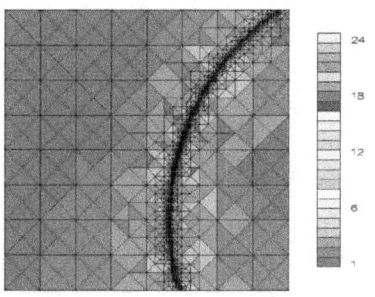

Figure 510: Example grid for the TYPEPARAM strategy with the asymmetric wave front problem.

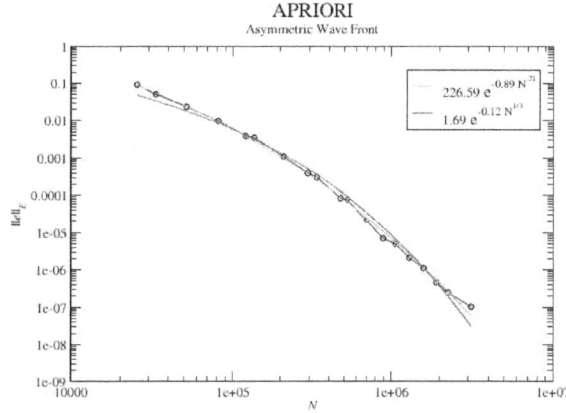

Figure 512: Log-Log plot of the convergence of the APRIORI strategy with the asymmetric wave front problem.

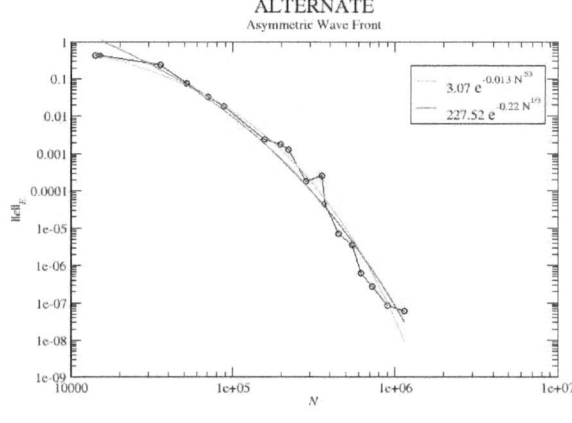

Figure 511: Log-Log plot of the convergence of the ALTERNATE strategy with the asymmetric wave front problem.

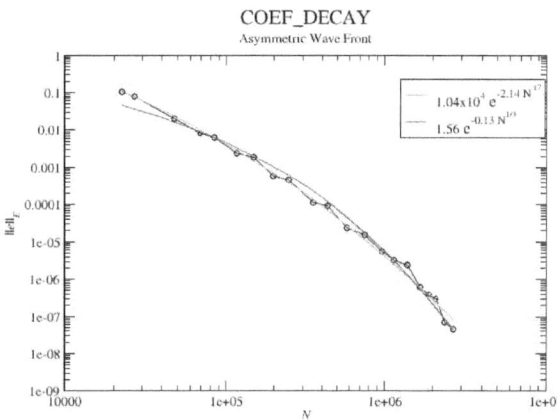

Figure 513: Log-Log plot of the convergence of the COEF-DECAY strategy with the asymmetric wave front problem.

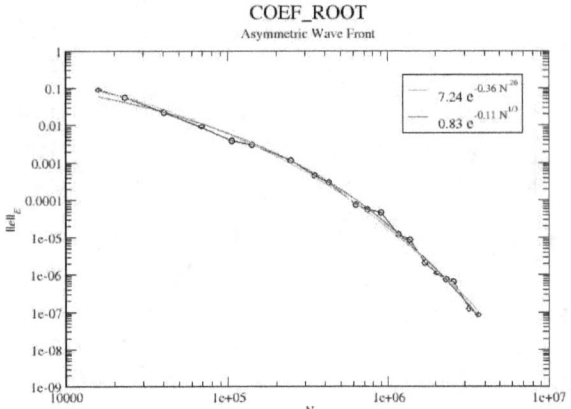

Figure 514: Log-Log plot of the convergence of the COEF-ROOT strategy with the asymmetric wave front problem.

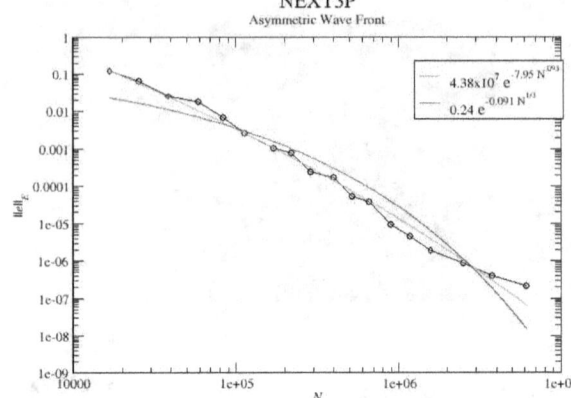

Figure 516: Log-Log plot of the convergence of the NEXT3P strategy with the asymmetric wave front problem.

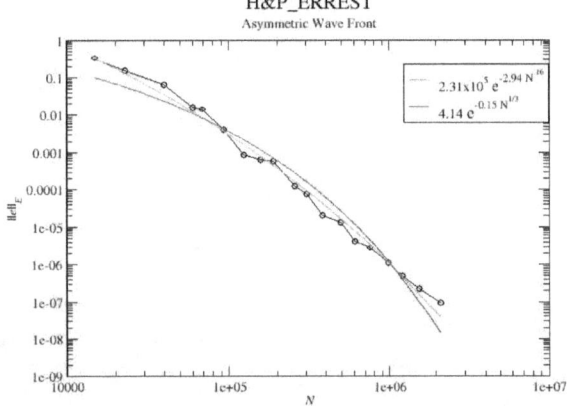

Figure 515: Log-Log plot of the convergence of the H&P-ERREST strategy with the asymmetric wave front problem.

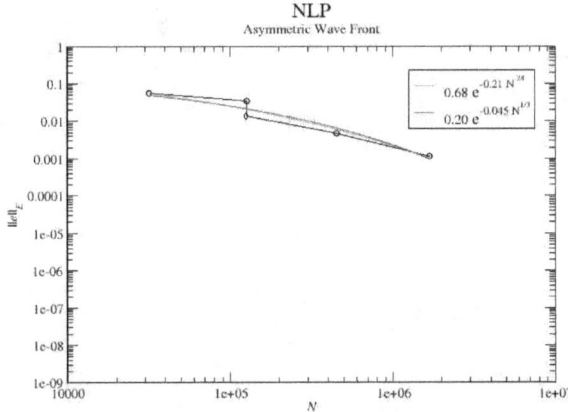

Figure 517: Log-Log plot of the convergence of the NLP strategy with the asymmetric wave front problem.

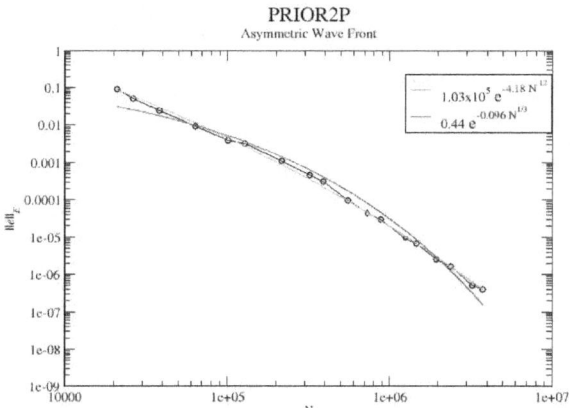

Figure 518: Log-Log plot of the convergence of the PRIOR2P strategy with the asymmetric wave front problem.

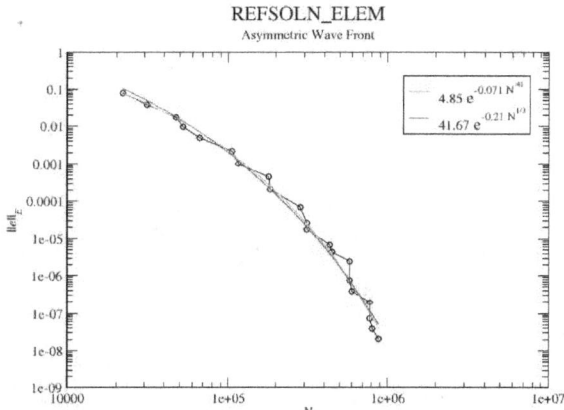

Figure 520: Log-Log plot of the convergence of the REFSOLN-ELEM strategy with the asymmetric wave front problem.

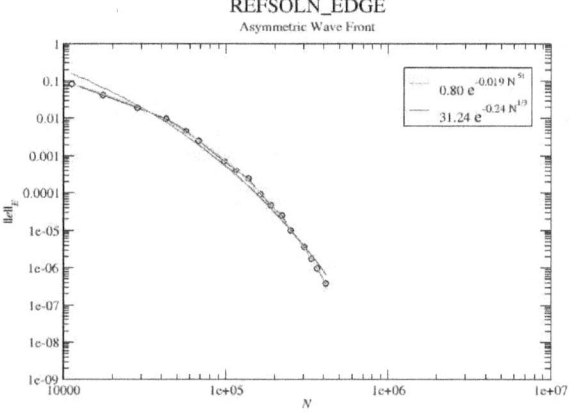

Figure 519: Log-Log plot of the convergence of the REFSOLN-EDGE strategy with the asymmetric wave front problem.

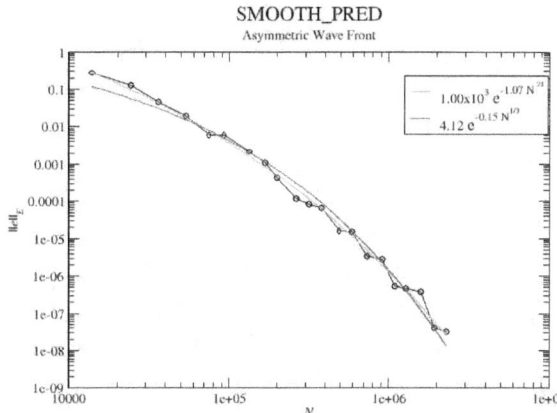

Figure 521: Log-Log plot of the convergence of the SMOOTH-PRED strategy with the asymmetric wave front problem.

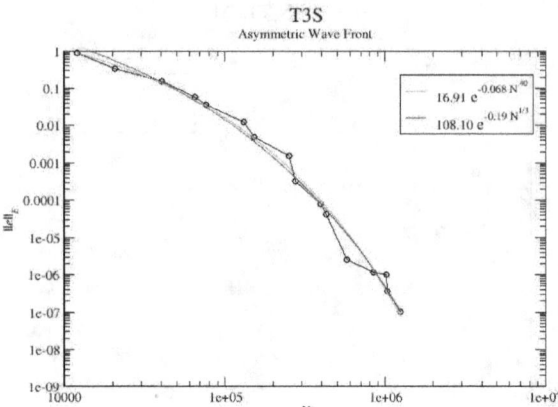

Figure 522: Log-Log plot of the convergence of the T3S strategy with the asymmetric wave front problem.

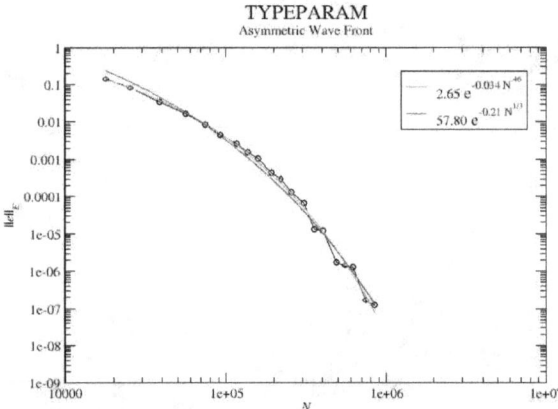

Figure 523: Log-Log plot of the convergence of the TYPEPARAM strategy with the asymmetric wave front problem.

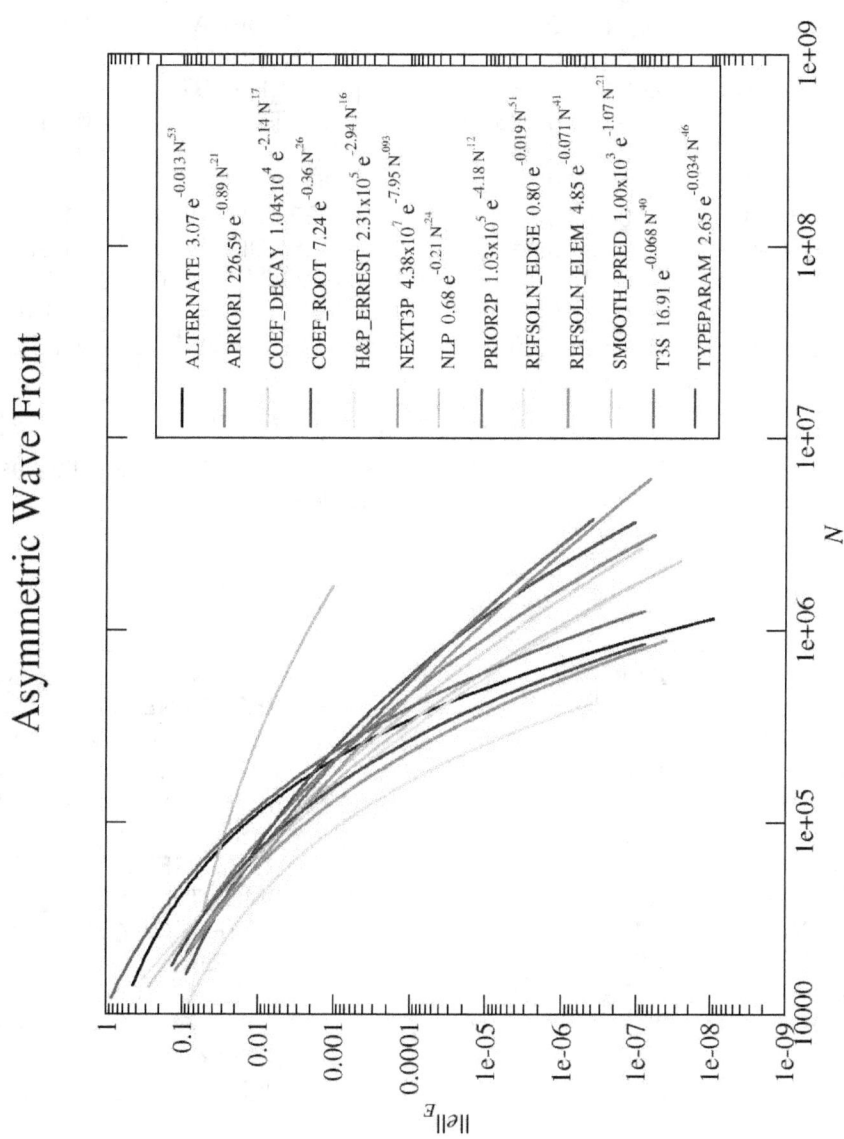

Figure 524: Log-Log plot of the convergence of all strategies with the asymmetric wave front problem.

strategy	A	B	C
ALTERNATE	3.07	-0.013	0.53
APRIORI	226.59	-0.89	0.21
COEFDECAY	1.04x10⁴	-2.14	0.17
COEFROOT	7.24	-0.36	0.26
H&PERREST	2.31x10⁵	-2.94	0.16
NEXT3P	4.38x10⁷	-7.95	0.093
NLP	0.68	-0.21	0.24
PRIOR2P	1.03x10⁶	-4.18	0.12
REFSOLNEDGE	0.80	-0.019	0.51
REFSOLNELEM	4.85	-0.071	0.41
SMOOTHPRED	1.00x10³	-1.07	0.21
T3S	16.91	-0.068	0.40
TYPEPARAM	2.65	-0.034	0.46

Table 69: Parameters of the least squares fit for $\|e_{hp}\|_E = Ae^{BN_{dof}^C}$ for the asymmetric wave front problem.

strategy	A	B
ALTERNATE	227.52	-0.22
APRIORI	1.69	-0.12
COEFDECAY	1.56	-0.13
COEFROOT	0.83	-0.11
H&PERREST	4.14	-0.15
NEXT3P	0.24	-0.091
NLP	0.20	-0.045
PRIOR2P	0.44	-0.096
REFSOLNEDGE	31.24	-0.24
REFSOLNELEM	41.67	-0.21
SMOOTHPRED	4.12	-0.15
T3S	108.10	-0.19
TYPEPARAM	57.80	-0.21

Table 71: Parameters of the least squares fit for $\|e_{hp}\|_E = Ae^{BN_{dof}^{1/3}}$ for the asymmetric wave front problem.

strategy	factor
REFSOLNEDGE	1.00
REFSOLNELEM	1.46
NEXT3P	1.52
H&PERREST	1.63
SMOOTHPRED	1.69
PRIOR2P	1.70
COEFDECAY	1.74
TYPEPARAM	1.77
COEFROOT	1.86
APRIORI	1.97
ALTERNATE	2.77
T3S	2.96
NLP	6.90

Table 70: Factor by which N is larger than the best strategy for the asymmetric wave front problem at low accuracy, 1.0×10^{-2}.

strategy	factor
REFSOLNEDGE	1.00
REFSOLNELEM	1.52
TYPEPARAM	1.64
ALTERNATE	1.88
T3S	2.42
H&PERREST	2.79
SMOOTHPRED	2.93
COEFDECAY	4.01
APRIORI	4.52
COEFROOT	5.99
NEXT3P	6.92
PRIOR2P	7.67
NLP	96.23

Table 72: Factor by which N is larger than the best strategy for the asymmetric wave front problem at high accuracy, 1.0×10^{-6}.

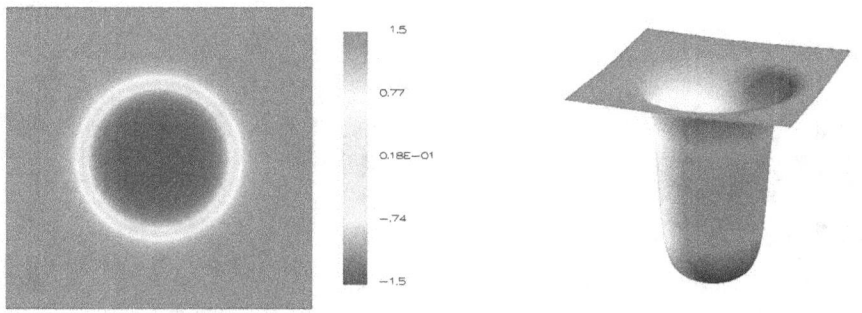

Figure 525: The solution of the singular well problem.

5.19 Singular Well

In the wavefront problems of the previous three sections, the center of the circle was placed outside the domain so the mild singularity at the center of the circle was not a factor. In the singular well problem, the center of the circle is placed at the center of the domain and the wavefront is relatively mild, effectively creating a well with a mild singularity at the center. $\alpha=50$, $(x_c, y_c)=(.5,.5)$, and $r_0=.25$. $\tau=10^{-3}$ for the grid images. For the APRIORI strategy, refine by h if the element touches the circle that defines the location of the wavefront and has degree at least 3 (chosen arbitrarily), or touches the center of the circle, and by p otherwise.

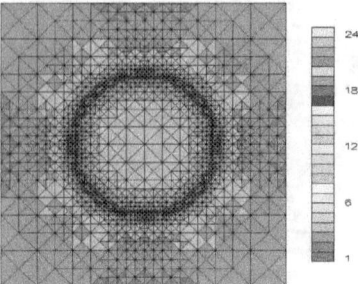

Figure 526: Example grid for the ALTERNATE strategy with the singular well problem.

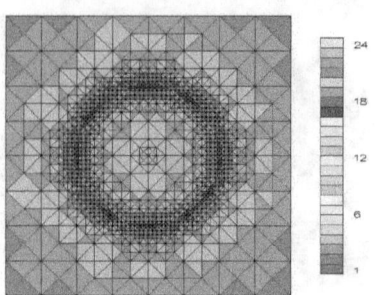

Figure 529: Example grid for the COEFROOT strategy with the singular well problem.

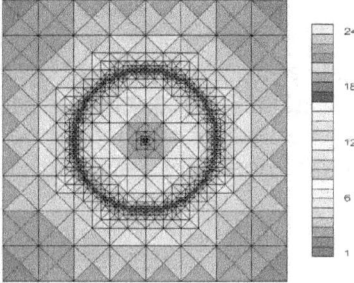

Figure 527: Example grid for the APRIORI strategy with the singular well problem.

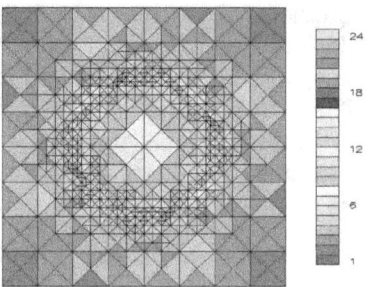

Figure 530: Example grid for the H&PERREST strategy with the singular well problem.

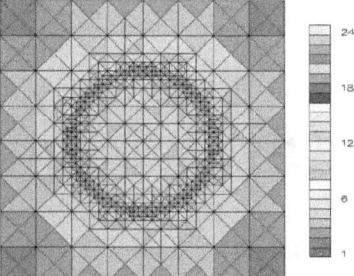

Figure 528: Example grid for the COEFDECAY strategy with the singular well problem.

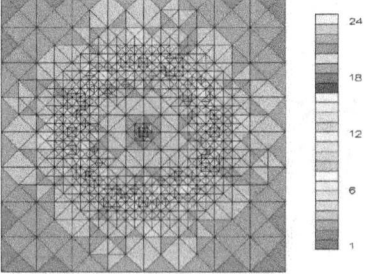

Figure 531: Example grid for the NEXT3P strategy with the singular well problem.

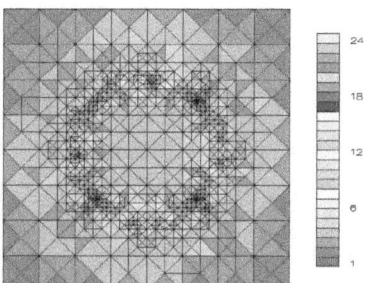

Figure 532: Example grid for the NLP strategy with the singular well problem.

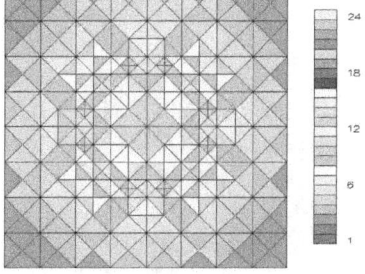

Figure 535: Example grid for the REFSOLNELEM strategy with the singular well problem.

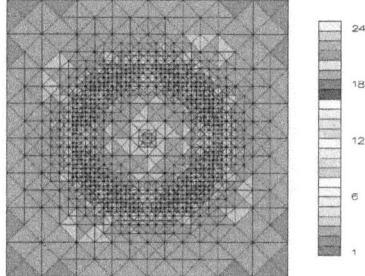

Figure 533: Example grid for the PRIOR2P strategy with the singular well problem.

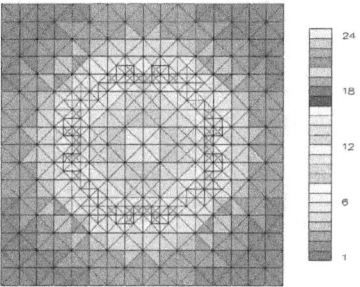

Figure 536: Example grid for the SMOOTHPRED strategy with the singular well problem.

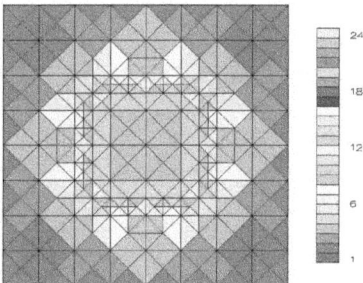

Figure 534: Example grid for the REFSOLNEDGE strategy with the singular well problem.

Figure 537: Example grid for the T3S strategy with the singular well problem.

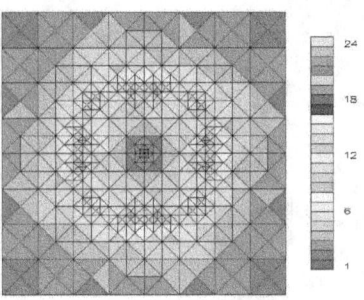

Figure 538: Example grid for the TYPEPARAM strategy with the singular well problem.

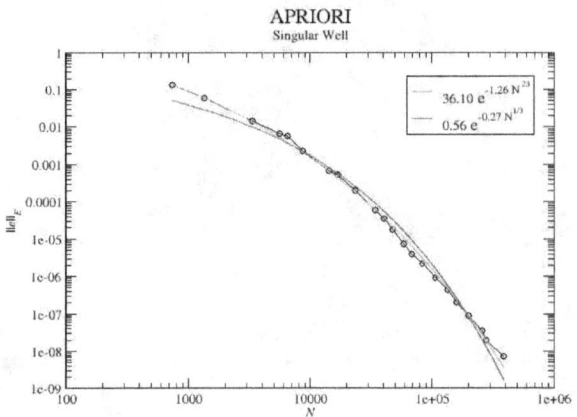

Figure 540: Log-Log plot of the convergence of the APRIORI strategy with the singular well problem.

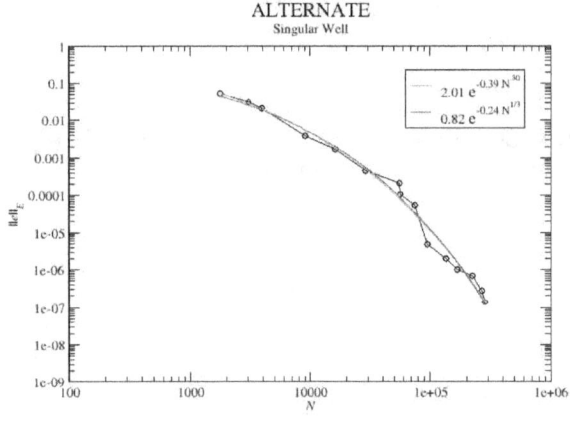

Figure 539: Log-Log plot of the convergence of the ALTERNATE strategy with the singular well problem.

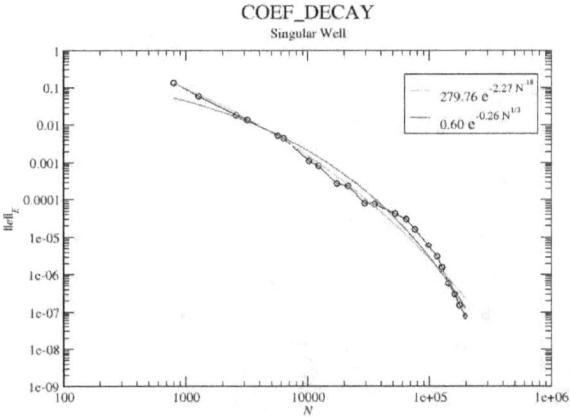

Figure 541: Log-Log plot of the convergence of the COEF-DECAY strategy with the singular well problem.

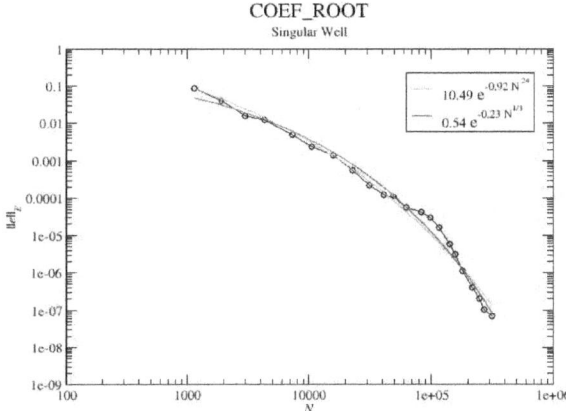

Figure 542: Log-Log plot of the convergence of the COEF-ROOT strategy with the singular well problem.

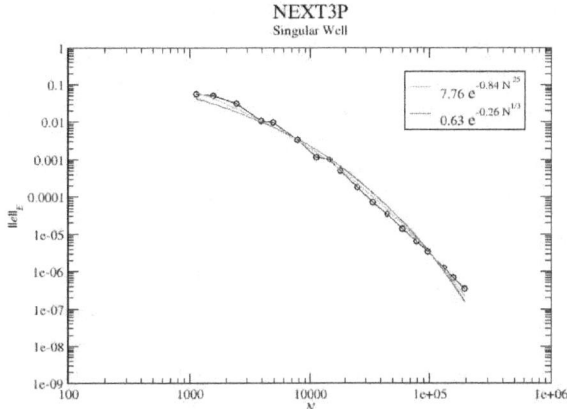

Figure 544: Log-Log plot of the convergence of the NEXT3P strategy with the singular well problem.

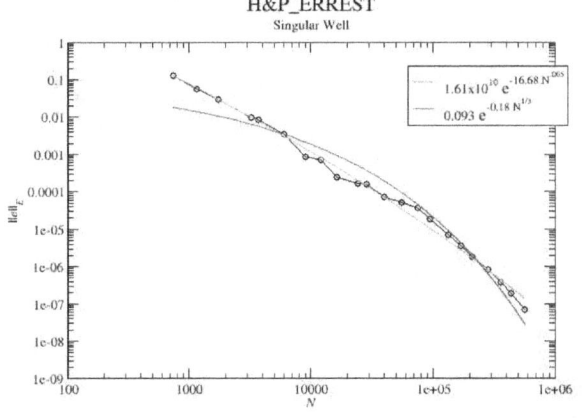

Figure 543: Log-Log plot of the convergence of the H&P-ERREST strategy with the singular well problem.

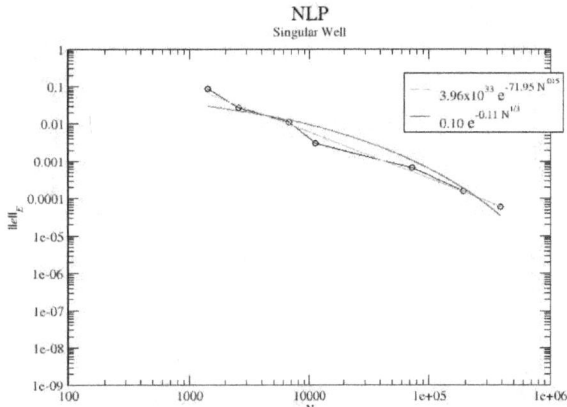

Figure 545: Log-Log plot of the convergence of the NLP strategy with the singular well problem.

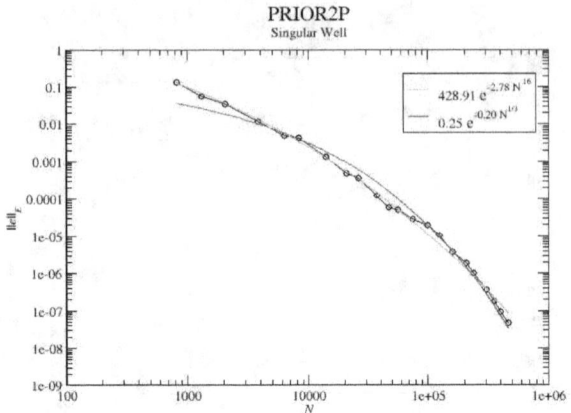

Figure 546: Log-Log plot of the convergence of the PRIOR2P strategy with the singular well problem.

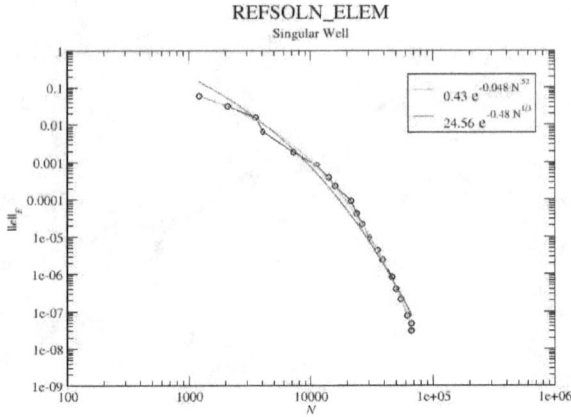

Figure 548: Log-Logplot of the convergence of the REFSOLN-ELEM strategy with the singular well problem.

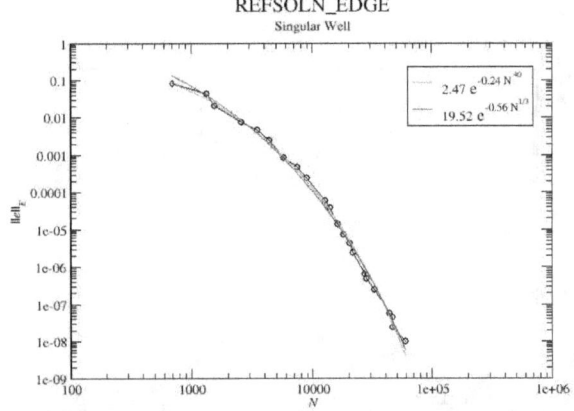

Figure 547: Log-Logplot of the convergence of the REFSOLN-EDGE strategy with the singular well problem.

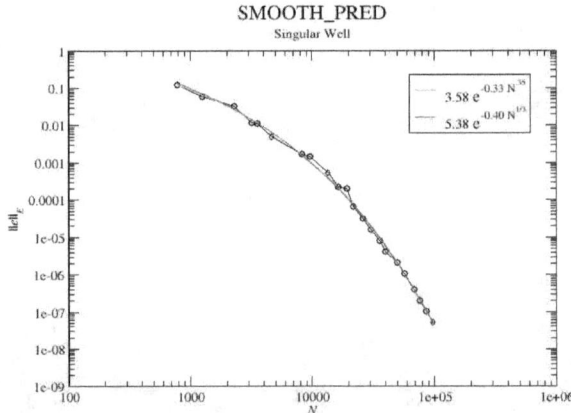

Figure 549: Log-Logplot of the convergence of the SMOOTH-PRED strategy with the singular well problem.

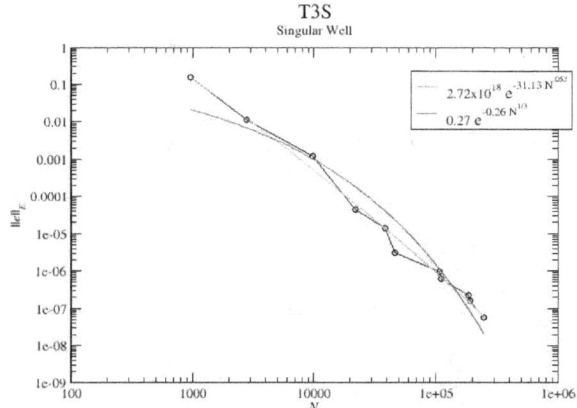

Figure 550: Log-Log plot of the convergence of the T3S strategy with the singular well problem.

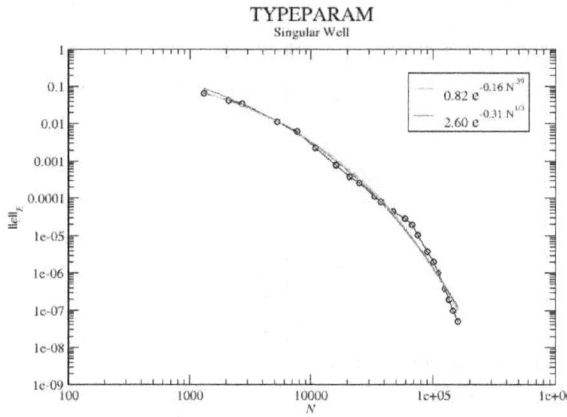

Figure 551: Log-Log plot of the convergence of the TYPEPARAM strategy with the singular well problem.

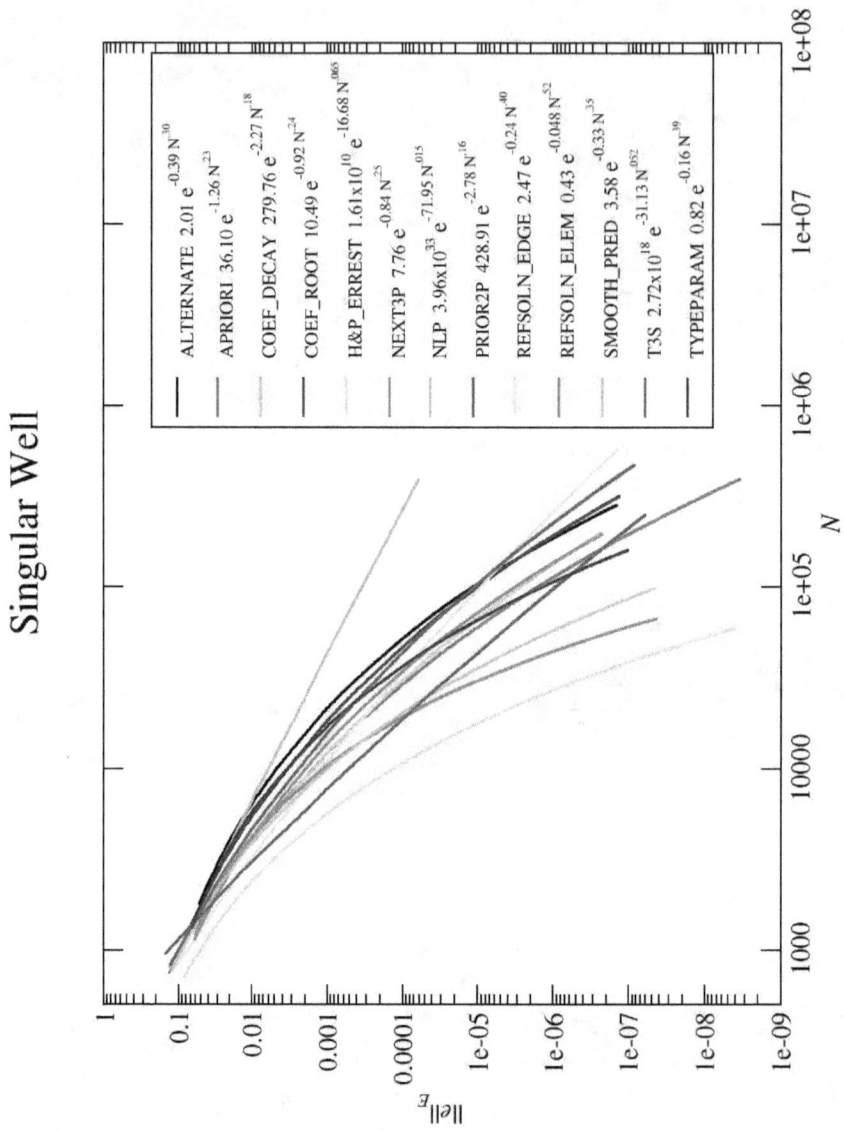

Figure 552: Log-Log plot of the convergence of all strategies with the singular well problem.

strategy	A	B	C
ALTERNATE	2.01	-0.39	0.30
APRIORI	36.10	-1.26	0.23
COEFDECAY	279.76	-2.27	0.18
COEFROOT	10.49	-0.92	0.24
H&PERREST	1.61×10^{10}	-16.68	0.065
NEXT3P	7.76	-0.84	0.25
NLP	3.96×10^{33}	-71.95	0.015
PRIOR2P	428.91	-2.78	0.16
REFSOLNEDGE	2.47	-0.24	0.40
REFSOLNELEM	0.43	-0.048	0.52
SMOOTHPRED	3.58	-0.33	0.35
T3S	2.72×10^{8}	-31.13	0.052
TYPEPARAM	0.82	-0.16	0.39

Table 73: Parameters of the least squares fit for $\|e_{hp}\|_E = Ae^{BN^{C}_{dof}}$ for the singular well problem.

strategy	A	B
ALTERNATE	0.82	-0.24
APRIORI	0.56	-0.27
COEFDECAY	0.60	-0.26
COEFROOT	0.54	-0.23
H&PERREST	0.093	-0.18
NEXT3P	0.63	-0.26
NLP	0.10	-0.11
PRIOR2P	0.25	-0.20
REFSOLNEDGE	19.52	-0.56
REFSOLNELEM	24.56	-0.48
SMOOTHPRED	5.38	-0.40
T3S	0.27	-0.26
TYPEPARAM	2.60	-0.31

Table 75: Parameters of the least squares fit for $\|e_{hp}\|_E = Ae^{BN^{1/3}_{dof}}$ for the singular well problem.

strategy	factor
REFSOLNEDGE	1.00
T3S	1.28
H&PERREST	1.36
SMOOTHPRED	1.62
COEFDECAY	1.67
APRIORI	1.69
NEXT3P	1.74
REFSOLNELEM	1.75
PRIOR2P	1.93
COEFROOT	2.28
TYPEPARAM	2.40
ALTERNATE	2.69
NLP	2.80

Table 74: Factor by which N is larger than the best strategy for the singular well problem at low accuracy, 1.0×10^{-2}.

strategy	factor
REFSOLNEDGE	1.00
REFSOLNELEM	1.64
SMOOTHPRED	2.13
T3S	3.62
TYPEPARAM	3.95
APRIORI	4.19
COEFDECAY	4.97
NEXT3P	5.04
ALTERNATE	6.90
COEFROOT	7.15
PRIOR2P	8.38
H&PERREST	9.63
NLP	326.76

Table 76: Factor by which N is larger than the best strategy for the singular well problem at high accuracy, 1.0×10^{-6}.

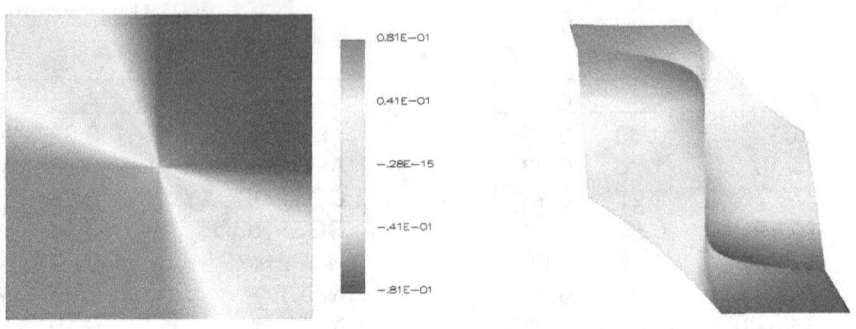

Figure 553: The solution of the intersecting interfaces problem.

5.20 Intersecting Interfaces

The intersecting interfaces problem has piecewise constant coefficients which create a very strong singularity at the center of the domain and discontinuous derivatives along the x and y axes. The boundary conditions are Dirichlet on the domain $(-1,1) \times (-1,1)$. For the grid images, $\tau = 5 \times 10^{-3}$. For the APRIORI strategy, refine by h if the element touches the origin and by p otherwise.

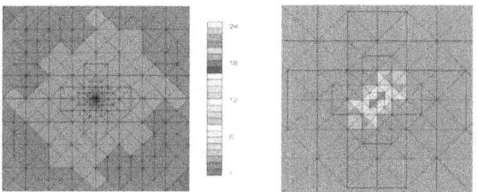

Figure 554: Example grid for the ALTERNATE strategy with the intersecting interfaces problem, including details at the singularity.

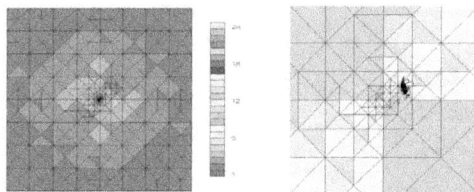

Figure 558: Example grid for the H&PERREST strategy with the intersecting interfaces problem, including details at the singularity.

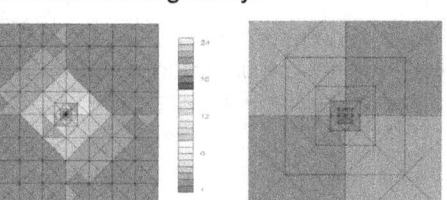

Figure 555: Example grid for the APRIORI strategy with the intersecting interfaces problem, including details at the singularity.

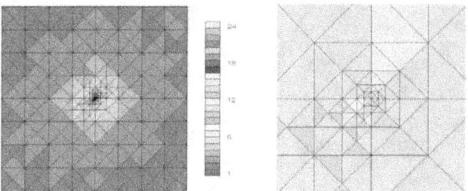

Figure 559: Example grid for the NEXT3P strategy with the intersecting interfaces problem, including details at the singularity.

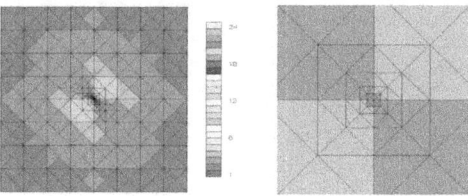

Figure 556: Example grid for the COEFDECAY strategy with the intersecting interfaces problem, including details at the singularity.

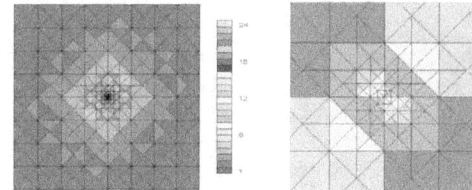

Figure 560: Example grid for the NLP strategy with the intersecting interfaces problem, including details at the singularity.

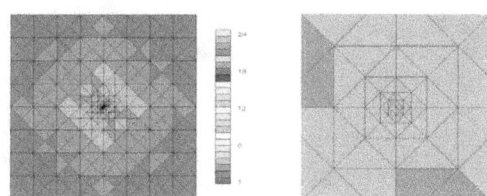

Figure 557: Example grid for the COEFROOT strategy with the intersecting interfaces problem, including details at the singularity.

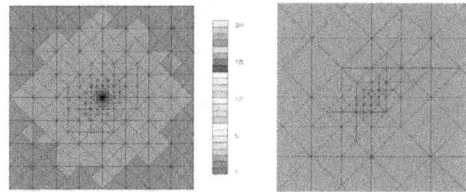

Figure 561: Example grid for the PRIOR2P strategy with the intersecting interfaces problem, including details at the singularity.

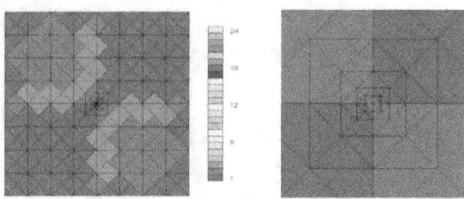

Figure 562: Example grid for the REFSOLNEDGE strategy with the intersecting interfaces problem, including details at the singularity.

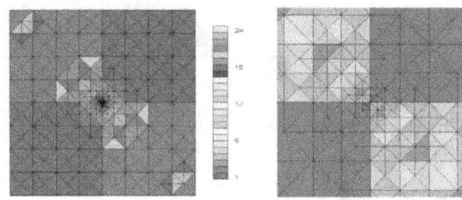

Figure 563: Example grid for the REFSOLNELEM strategy with the intersecting interfaces problem, including details at the singularity.

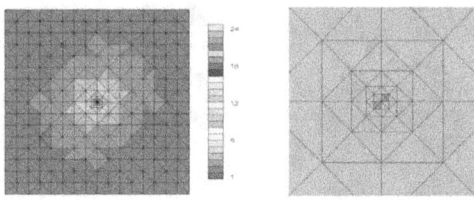

Figure 564: Example grid for the SMOOTHPRED strategy with the intersecting interfaces problem, including details at the singularity.

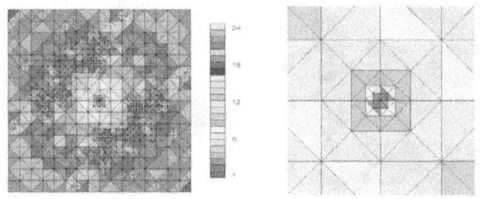

Figure 565: Example grid for the T3S strategy with the intersecting interfaces problem, including details at the singularity.

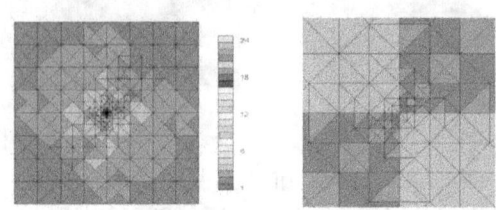

Figure 566: Example grid for the TYPEPARAM strategy with the intersecting interfaces problem, including details at the singularity.

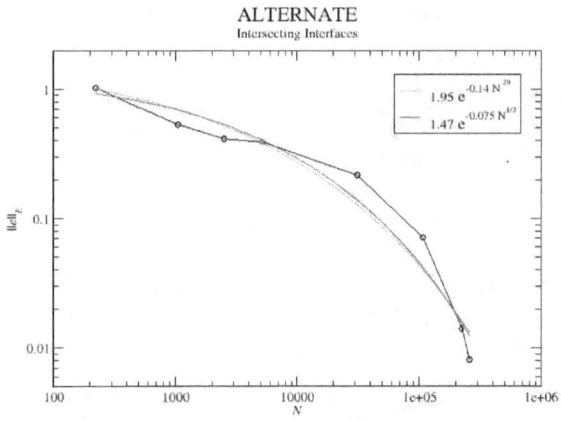

Figure 567: Log-Log plot of the convergence of the ALTERNATE strategy with the intersecting interfaces problem.

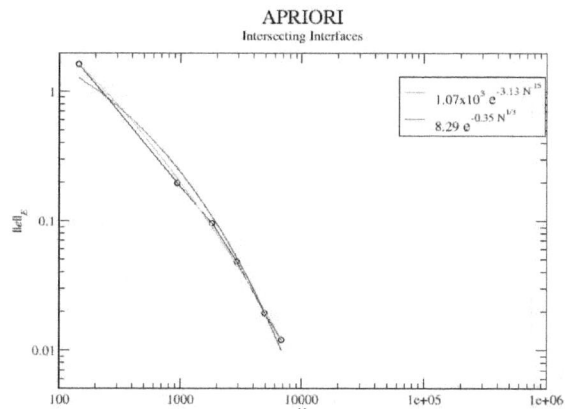
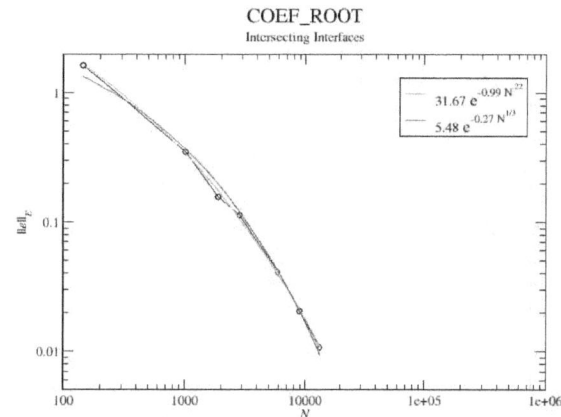

Figure 568: Log-Log plot of the convergence of the APRIORI strategy with the intersecting interfaces problem.

Figure 570: Log-Log plot of the convergence of the COEF-ROOT strategy with the intersecting interfaces problem.

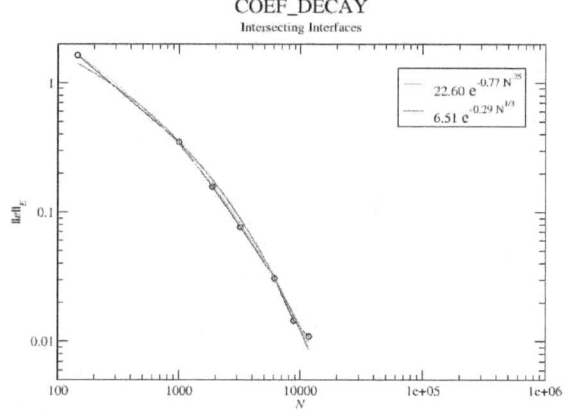
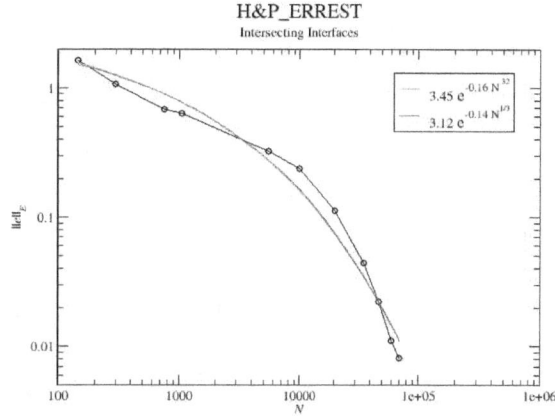

Figure 569: Log-Log plot of the convergence of the COEF-DECAY strategy with the intersecting interfaces problem.

Figure 571: Log-Log plot of the convergence of the H&P-ERREST strategy with the intersecting interfaces problem.

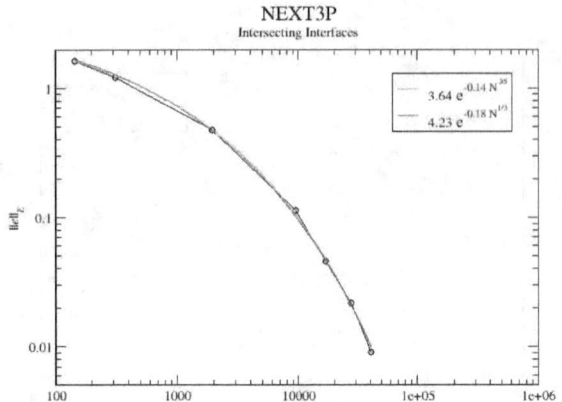
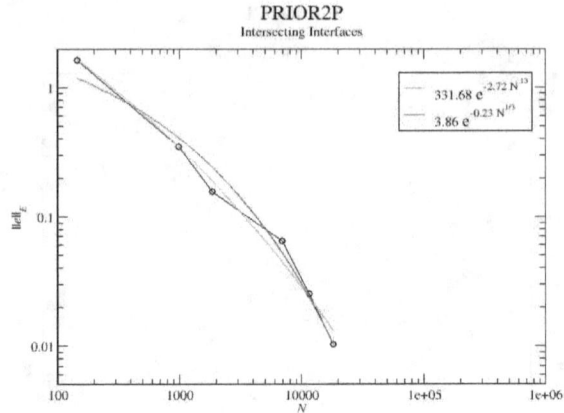

Figure 572: Log-Logplot of the convergence of the NEXT3P strategy with the intersecting interfaces problem.

Figure 574: Log-Logplot of the convergence of the PRIOR2P strategy with the intersecting interfaces problem.

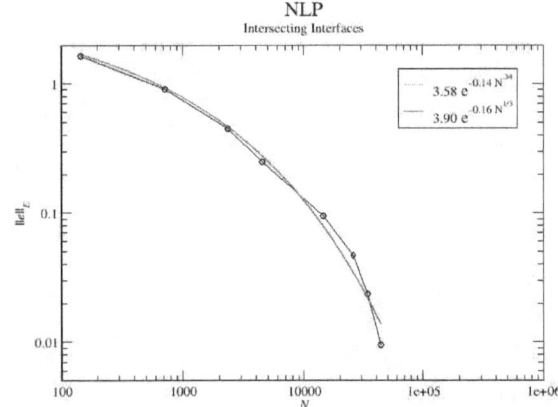
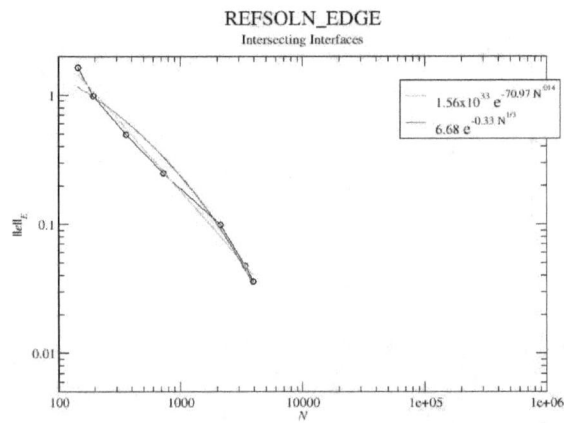

Figure 573: Log-Logplot of the convergence of the NLP strategy with the intersecting interfaces problem.

Figure 575: Log-Logplot of the convergence of the REFSOLN-EDGE strategy with the intersecting interfaces problem.

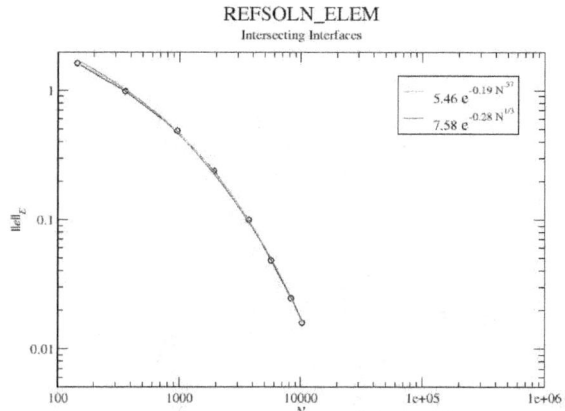

Figure 576: Log-Log plot of the convergence of the REFSOLN-ELEM strategy with the intersecting interfaces problem.

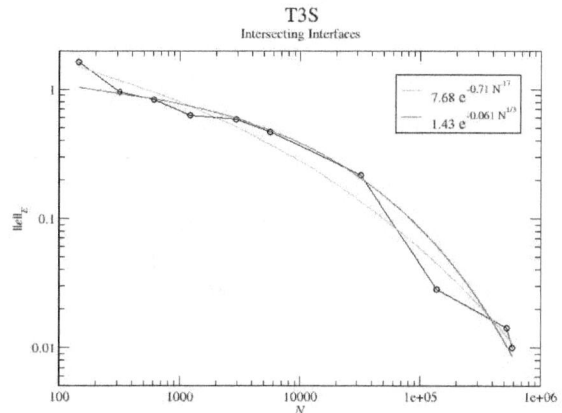

Figure 578: Log-Log plot of the convergence of the T3S strategy with the intersecting interfaces problem.

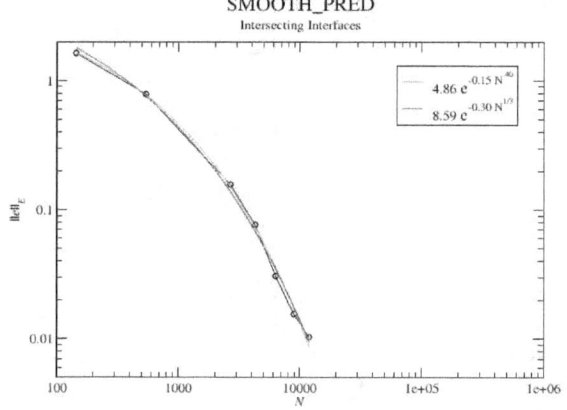

Figure 577: Log-Log plot of the convergence of the SMOOTH-PRED strategy with the intersecting interfaces problem.

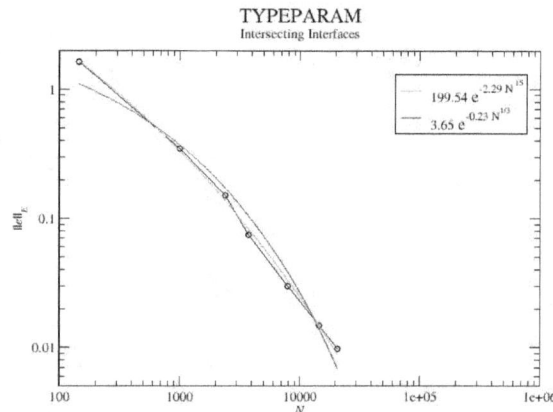

Figure 579: Log-Log plot of the convergence of the TYPEPARAM strategy with the intersecting interfaces problem.

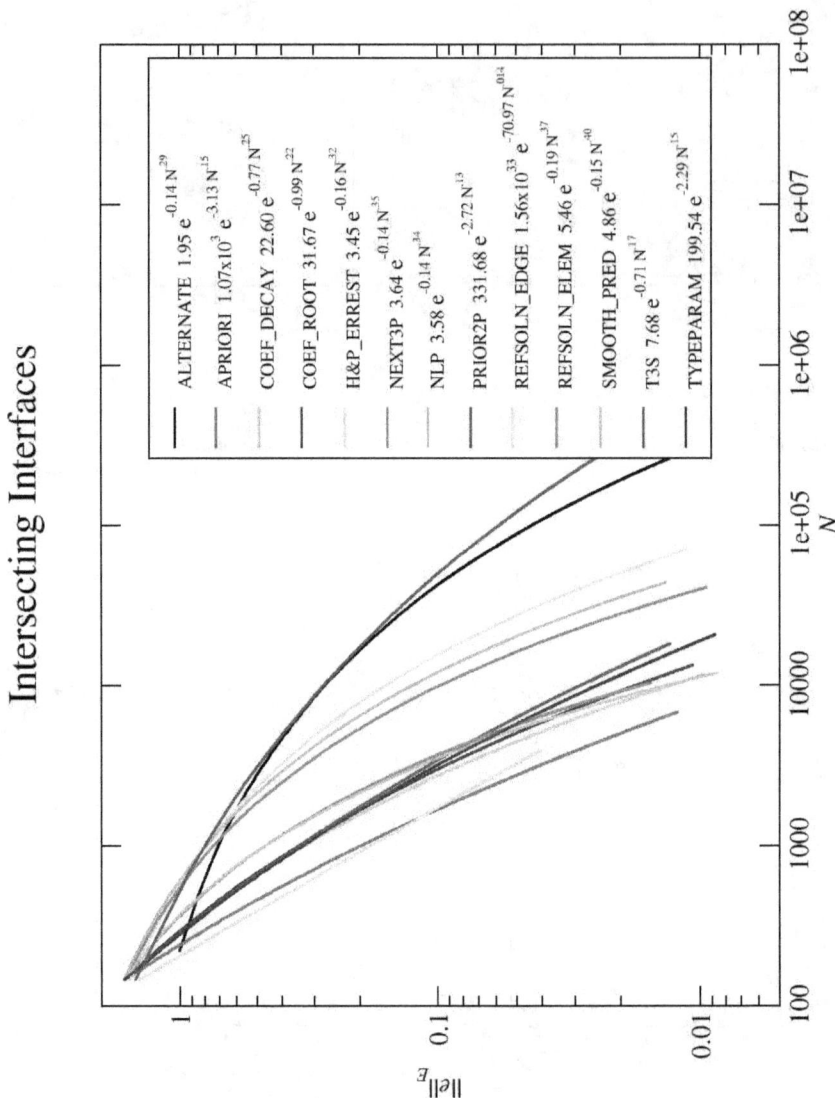

Figure 580: Log-Log plot of the convergence of all strategies with the intersecting interfaces problem.

strategy	A	B	C
ALTERNATE	1.95	-0.14	0.29
APRIORI	1.07x10³	-3.13	0.15
COEFDECAY	22.60	-0.77	0.25
COEFROOT	31.67	-0.99	0.22
H&PERREST	3.45	-0.16	0.32
NEXT3P	3.64	-0.14	0.35
NLP	3.58	-0.14	0.34
PRIOR2P	331.68	-2.72	0.13
REFSOLNEDGE	1.56x10³³	-70.97	0.014
REFSOLNELEM	5.46	-0.19	0.37
SMOOTHPRED	4.86	-0.15	0.40
T3S	7.68	-0.71	0.17
TYPEPARAM	199.54	-2.29	0.15

Table 77: Parameters of the least squares fit for $\|e_{hp}\|_E = A e^{BN_{dof}^C}$ for the intersecting interfaces problem.

strategy	A	B
ALTERNATE	1.47	-0.075
APRIORI	8.29	-0.35
COEFDECAY	6.51	-0.29
COEFROOT	5.48	-0.27
H&PERREST	3.12	-0.14
NEXT3P	4.23	-0.18
NLP	3.90	-0.16
PRIOR2P	3.86	-0.23
REFSOLNEDGE	6.68	-0.33
REFSOLNELEM	7.58	-0.28
SMOOTHPRED	8.59	-0.30
T3S	1.43	-0.061
TYPEPARAM	3.65	-0.23

Table 79: Parameters of the least squares fit for $\|e_{hp}\|_E = A e^{BN_{dof}^{1/3}}$ for the intersecting interfaces problem.

strategy	factor
REFSOLNEDGE	1.00
APRIORI	1.08
COEFDECAY	1.71
COEFROOT	1.83
TYPEPARAM	1.83
PRIOR2P	1.92
SMOOTHPRED	2.35
REFSOLNELEM	2.41
NEXT3P	5.88
NLP	7.08
H&PERREST	8.58
ALTERNATE	18.30
T3S	19.15

Table 78: Factor by which N is larger than the best strategy for the intersecting interfaces problem at low accuracy, 2.0×10^{-1}.

strategy	factor
APRIORI	1.00
REFSOLNEDGE	1.47
COEFDECAY	1.56
SMOOTHPRED	1.67
REFSOLNELEM	1.83
COEFROOT	1.84
TYPEPARAM	2.32
PRIOR2P	2.63
NEXT3P	5.57
NLP	7.07
H&PERREST	10.01
ALTERNATE	38.16
T3S	65.48

Table 80: Factor by which N is larger than the best strategy for the intersecting interfaces problem at high accuracy, 2.0×10^{-2}.

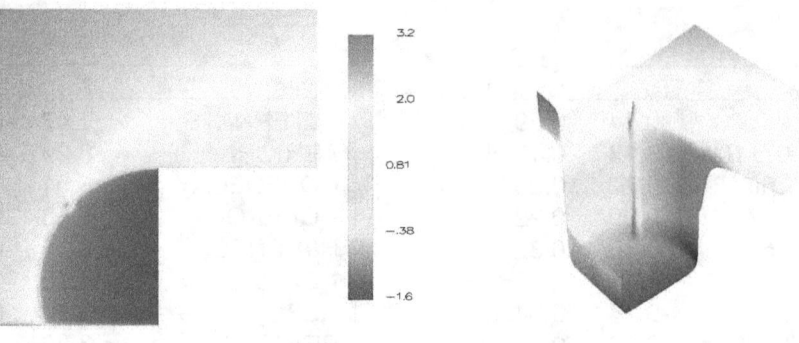

Figure 581: The solution of the multiple difficulties problem.

5.21 Multiple Difficulties

The multiple difficulties problem combines several of the difficulties of the other problems into a single problem. It contains a reentrant corner point singularity, wavefront, peak and boundary layer. For the selected parameters, the peak falls on the wavefront, and the wavefront intersects the boundary layer and point singularity. The parameters are:

- reentrant corner $\omega = 3\pi/2$
- center of circle for wavefront $(0, -3/4)$
- radius of circle for wavefront $3/4$
- strength of wavefront $\alpha = 200$
- center of peak $(\sqrt{5}/4, -1/4)$
- strength of peak $\alpha = 1000$
- strength of boundary layer $\varrho = 1/100$

For the grid images, $\tau = 10^{-2}$. The APRIORI method refines by h in the same cases as it did in the individual problems.

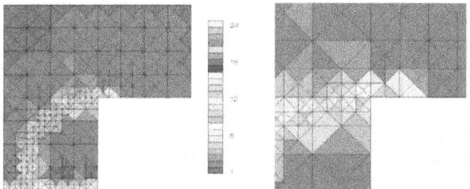

Figure 582: Example grid for the ALTERNATE strategy with the multiple difficulties problem, including details at the singularity.

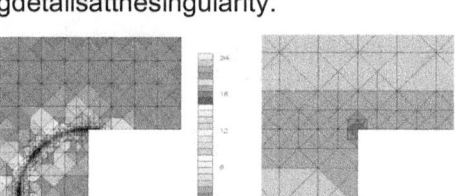

Figure 583: Example grid for the APRIORI strategy with the multiple difficulties problem, including details at the singularity.

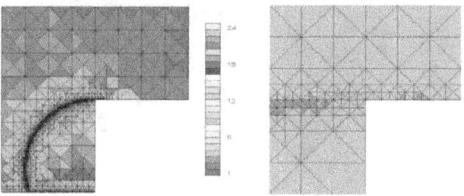

Figure 584: Example grid for the COEFDECAY strategy with the multiple difficulties problem, including details at the singularity.

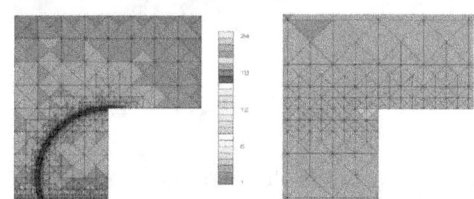

Figure 585: Example grid for the COEFROOT strategy with the multiple difficulties problem, including details at the singularity.

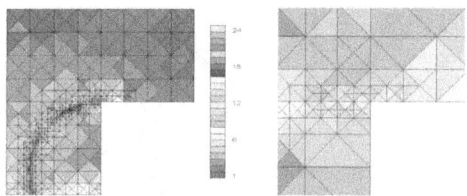

Figure 586: Example grid for the H&PERREST strategy with the multiple difficulties problem, including details at the singularity.

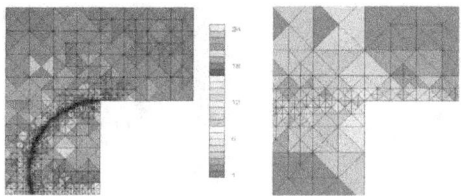

Figure 587: Example grid for the NEXT3P strategy with the multiple difficulties problem, including details at the singularity.

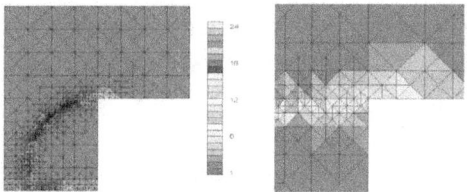

Figure 588: Example grid for the NLP strategy with the multiple difficulties problem, including details at the singularity.

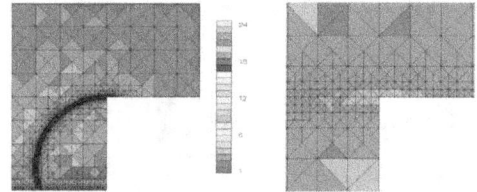

Figure 589: Example grid for the PRIOR2P strategy with the multiple difficulties problem, including details at the singularity.

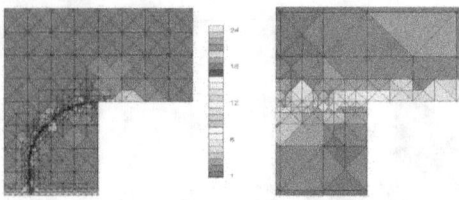

Figure 590: Example grid for the REFSOLNEDGE strategy with the multiple difficulties problem, including details at the singularity.

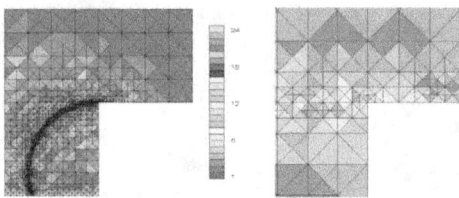

Figure 591: Example grid for the REFSOLNELEM strategy with the multiple difficulties problem, including details at the singularity.

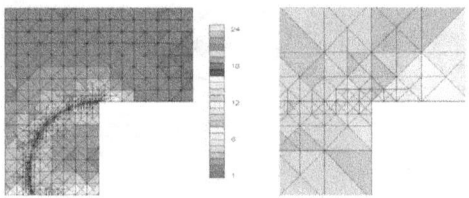

Figure 592: Example grid for the SMOOTHPRED strategy with the multiple difficulties problem, including details at the singularity.

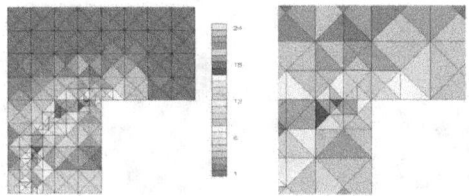

Figure 593: Example grid for the T3S strategy with the multiple difficulties problem, including details at the singularity.

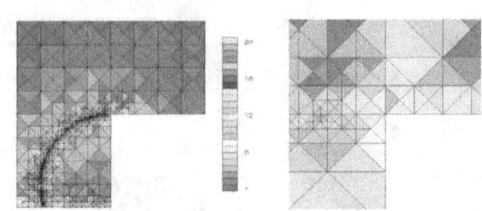

Figure 594: Example grid for the TYPEPARAM strategy with the multiple difficulties problem, including details at the singularity.

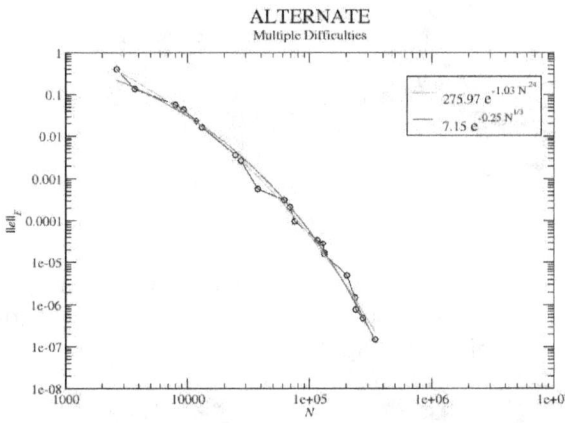

Figure 595: Log-Log plot of the convergence of the ALTERNATE strategy with the multiple difficulties problem.

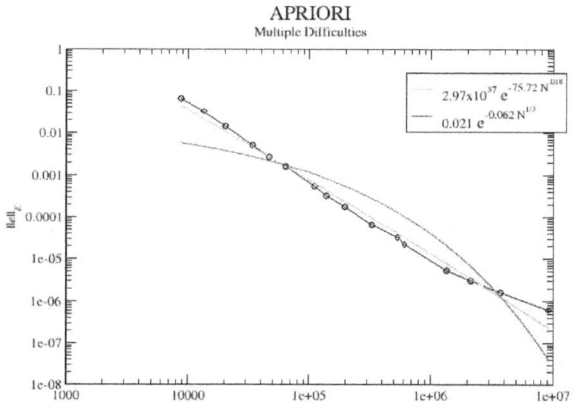

Figure 596: Log-Log plot of the convergence of the APRIORI strategy with the multiple difficulties problem.

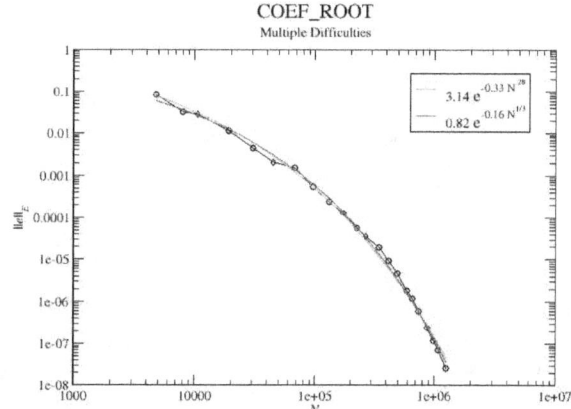

Figure 598: Log-Log plot of the convergence of the COEF-ROOT strategy with the multiple difficulties problem.

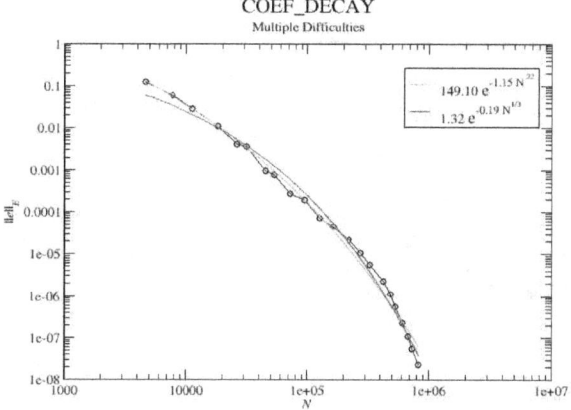

Figure 597: Log-Log plot of the convergence of the COEF-DECAY strategy with the multiple difficulties problem.

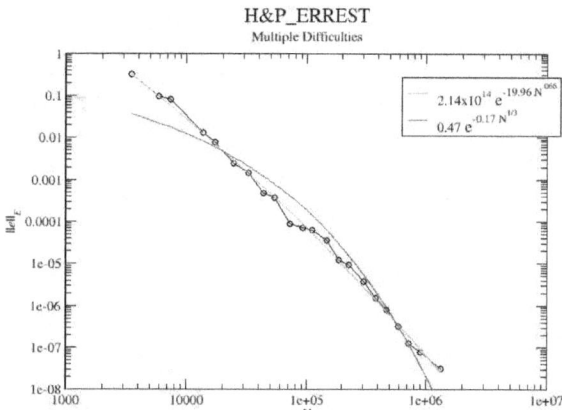

Figure 599: Log-Log plot of the convergence of the H&P-ERREST strategy with the multiple difficulties problem.

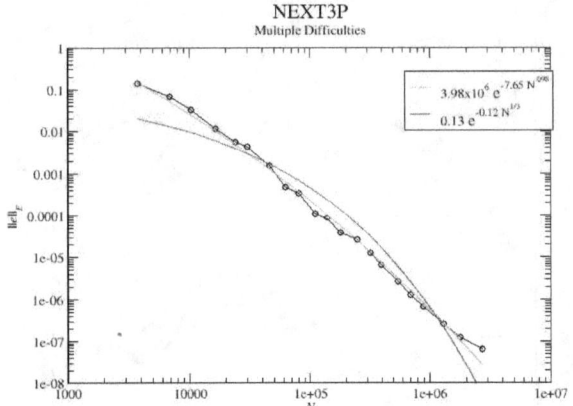

Figure 600: Log-Log plot of the convergence of the NEXT3P strategy with the multiple difficulties problem.

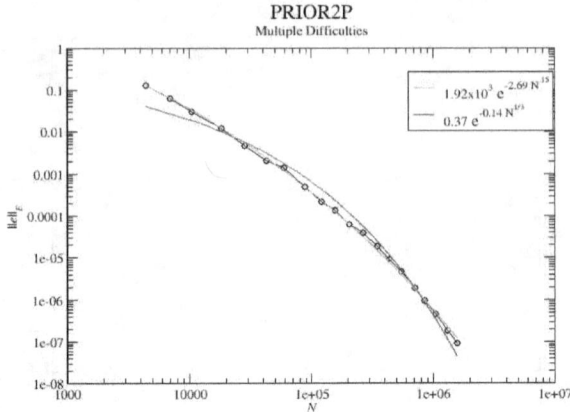

Figure 602: Log-Log plot of the convergence of the PRIOR2P strategy with the multiple difficulties problem.

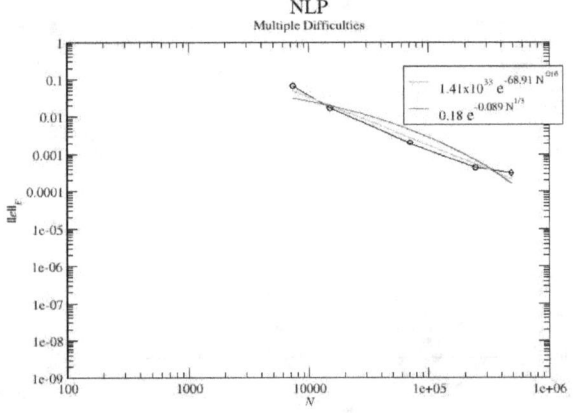

Figure 601: Log-Log plot of the convergence of the NLP strategy with the multiple difficulties problem.

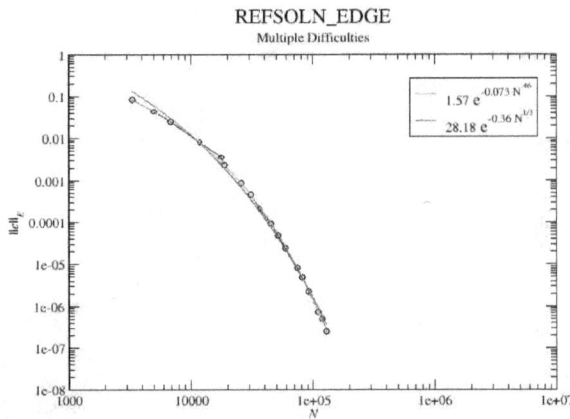

Figure 603: Log-Log plot of the convergence of the REFSOLN-EDGE strategy with the multiple difficulties problem.

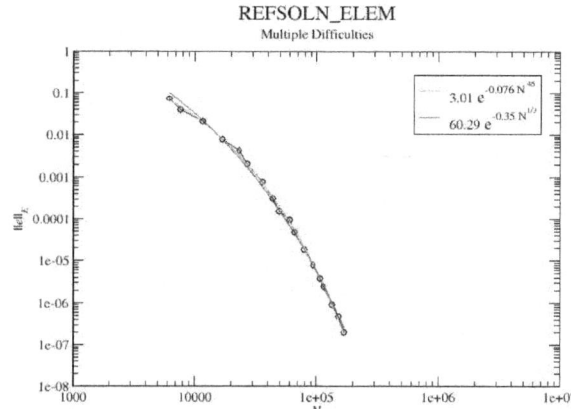

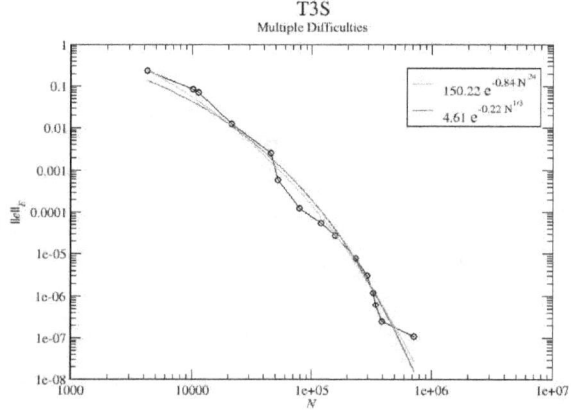

Figure 604: Log-Log plot of the convergence of the REFSOLN-ELEM strategy with the multiple difficulties problem.

Figure 606: Log-Log plot of the convergence of the T3S strategy with the multiple difficulties problem.

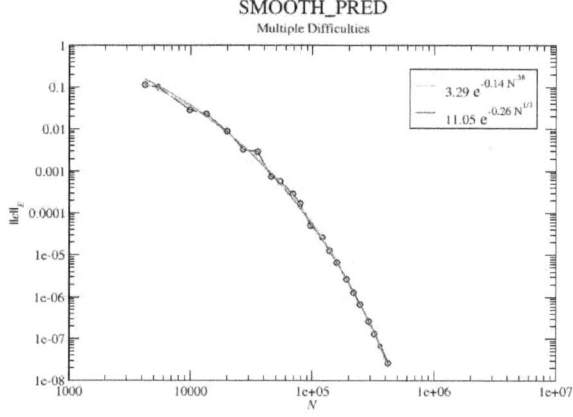

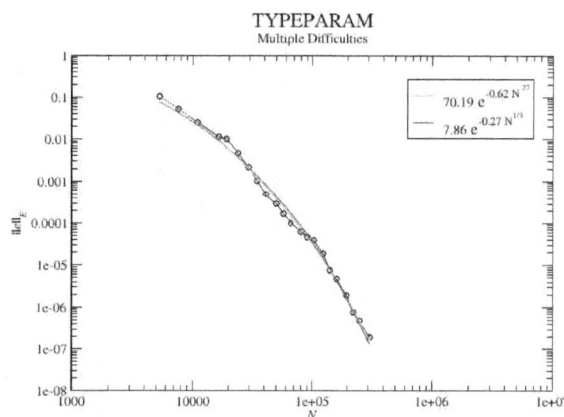

Figure 605: Log-Log plot of the convergence of the SMOOTH-PRED strategy with the multiple difficulties problem.

Figure 607: Log-Log plot of the convergence of the TYPEPARAM strategy with the multiple difficulties problem.

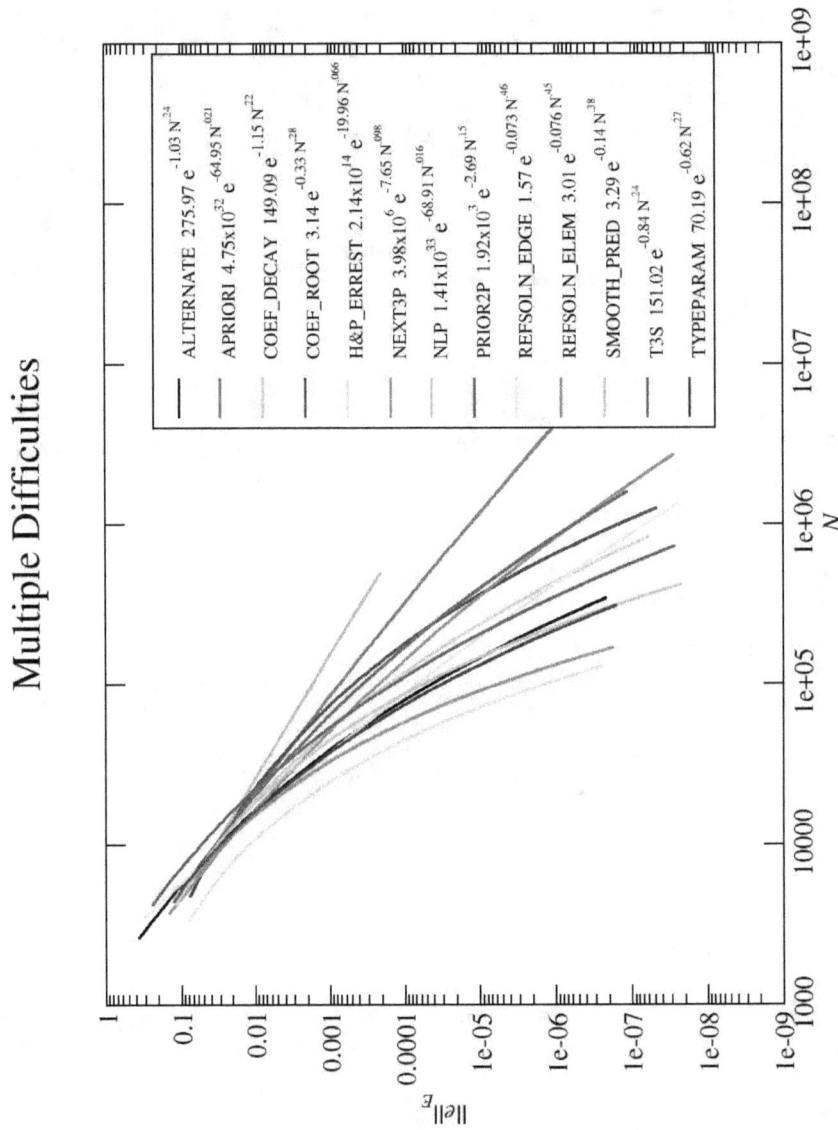

Figure 608: Log-Log plot of the convergence of all strategies with the multiple difficulties problem.

strategy	A	B	C
ALTERNATE	275.97	-1.03	0.24
APRIORI	2.97×10^{37}	-75.72	0.018
COEFDECAY	149.10	-1.15	0.22
COEFROOT	3.14	-0.33	0.28
H&PERREST	2.14×10^{14}	-19.96	0.066
NEXT3P	3.98×10^{6}	-7.65	0.098
NLP	1.41×10^{33}	-68.91	0.016
PRIOR2P	1.92×10^{3}	-2.69	0.15
REFSOLNEDGE	1.57	-0.073	0.46
REFSOLNELEM	3.01	-0.076	0.45
SMOOTHPRED	3.29	-0.14	0.38
T3S	150.22	-0.84	0.24
TYPEPARAM	70.19	-0.62	0.27

Table 81: Parameters of the least squares fit for $\|e_{hp}\|_E = A e^{B N_{dof}^C}$ for the multiple difficulties problem.

strategy	A	B
ALTERNATE	7.15	-0.25
APRIORI	0.021	-0.062
COEFDECAY	1.32	-0.19
COEFROOT	0.82	-0.16
H&PERREST	0.47	-0.17
NEXT3P	0.13	-0.12
NLP	0.18	-0.089
PRIOR2P	0.37	-0.14
REFSOLNEDGE	28.18	-0.36
REFSOLNELEM	60.29	-0.35
SMOOTHPRED	11.05	-0.26
T3S	4.61	-0.22
TYPEPARAM	7.86	-0.27

Table 83: Parameters of the least squares fit for $\|e_{hp}\|_E = A e^{B N_{dof}^{1/3}}$ for the multiple difficulties problem.

strategy	factor
REFSOLNEDGE	1.00
H&PERREST	1.40
REFSOLNELEM	1.43
TYPEPARAM	1.49
ALTERNATE	1.52
NEXT3P	1.53
SMOOTHPRED	1.71
COEFDECAY	1.77
PRIOR2P	1.90
APRIORI	1.98
T3S	2.05
COEFROOT	2.09
NLP	2.40

Table 82: Factor by which N is larger than the best strategy for the multiple difficulties problem at low accuracy, 1.0×10^{-2}.

strategy	factor
REFSOLNEDGE	1.00
REFSOLNELEM	1.26
TYPEPARAM	2.07
SMOOTHPRED	2.18
ALTERNATE	2.37
T3S	3.34
H&PERREST	4.02
COEFDECAY	4.13
COEFROOT	6.20
NEXT3P	7.82
PRIOR2P	8.00
APRIORI	39.60
NLP	224.68

Table 84: Factor by which N is larger than the best strategy for the multiple difficulties problem at high accuracy, 1.0×10^{-6}.

5.22 Computation Time

In the previous sections we presented performance results for the hp-adaptive strategies on a number of test problems in terms of error vs. the number of degrees of freedom. It would be interesting to also see a comparison in terms of error vs. computation time. However, we do not believe we could perform a fair comparison of that nature at this time for a number of reasons, not the least of which is that the implementation of the strategies in PHAML emphasized correctness of the method and was not tuned for optimal performance. Nevertheless, to satisfy one's curiosity about computation time, we present timing results for one problem, the mild peak problem at a tolerance of 10^{-6}. These times should not be taken too seriously; they should only be viewed as a rough estimate of the relative time required by each of the strategies.

These computations were performed in single user mode on a single core of a Dell Latitude D630 with the Intel Core 2 Duo processor T7700 operating under the CentOS 5.5 distribution of Linux with the 2.6.18 kernel. PHAML Version 1.8.1 was compiled with the Intel Fortran 95 compiler Version 11.1.072 using -O for optimization.

The results are given in Table 85. The first column gives the total wall clock time (in seconds) spent in refinement. There is some variation in the number of times each strategy went through the refine/solve loop making it difficult to compare the times spent in a single refinement phase of the loop using only these numbers. The second column gives the number of refine/solve loops, and the third gives the quotient of the first two columns to obtain the average time spent in a refinement phase. These figures show pretty much what one would expect *apriori*. Most of the strategies use between 0.035 and 0.111 seconds per refinement phase, which, due to the considerations above, should be considered approximately equal in this context, roughly .07 seconds. The H&P -ERREST strategy takes about twice as long, which makes sense because it computes two error indicators instead of one. The NEXT3P strategy takes about ten times longer, which makes sense because, not only is it computing three error indicators, but those error indicators are more expensive than the basic error indicator because they use a higher polynomial degree. The two reference solution strategies are roughly equal and take much longer than most strategies because they solve the expensive reference solution. Finally, NLP is extremely expensive, taking about 5000 times as long as the typical strategy because it has to solve the optimization problem.

5.23 Summary and Observations

In this section, we summarize the results in Sections 5.1–5.21 to examine the relative performance of the strategies in different situations. The test problems are grouped into six categories: easy problems, hard problems, and singular problems at low accuracy and high accuracy. We present the comparisons in two forms.

Tables 86–91 give a straight-forward ranking of the strategies for each problem based on the 3-parameter least squares fit. The four best strategies for each problem are highlighted in green, and the four worst in red to make it easy to see which strategies are consistently good or bad in a given category.

strategy	total time in refinement (s.)	number of ref/solve loops	average time per refinement (s./loop)
ALTERNATE	1.55	16	0.097
APRIORI	0.95	27	0.035
COEFDECAY	0.94	11	0.085
COEFROOT	0.88	12	0.073
H&PERREST	1.44	11	0.131
NEXT3P	7.09	11	0.645
NLP	3969.16	13	305.320
PRIOR2P	1.33	12	0.111
REFSOLNEDGE	29.38	19	1.546
REFSOLNELEM	20.01	12	1.668
SMOOTHPRED	1.03	11	0.094
T3S	0.38	8	0.048
TYPEPARAM	1.08	15	0.072

Table 85: Wall clock time for the refinement phases of the solution of the mild peak problem with $\tau = 10^{-6}$, the number of refine/solve loops, and the average time for a refinement phase of the loops.

strategy	analytic	mild peak	mild boundary layer	mild oscillatory	mild wave front
ALTERNATE	11	13	11	12	12
APRIORI	4	3	13	1	6
COEFDECAY	6	12	6	9	5
COEFROOT	7	9	8	13	10
H&PERREST	9	6	1	8	2
NEXT3P	2	5	4	6	4
NLP	12	7	7	10	11
PRIOR2P	8	10	5	11	7
REFSOLNEDGE	1	1	3	3	1
REFSOLNELEM	5	2	9	2	3
SMOOTHPRED	10	11	10	7	8
T3S	13	8	12	5	13
TYPEPARAM	3	4	2	4	9

Table 86: Low accuracy ranking of each strategy for easy problems.

strategy	analytic	mild peak	mild boundary layer	mild oscillatory	mild wave front
ALTERNATE	12	13	12	12	13
APRIORI	1	2	13	1	8
COEFDECAY	8	12	5	8	7
COEFROOT	6	10	8	9	9
H&PERREST	13	8	11	13	11
NEXT3P	9	7	9	10	10
NLP	7	9	6	7	4
PRIOR2P	10	11	10	11	12
REFSOLNEDGE	4	4	2	3	2
REFSOLNELEM	3	1	4	2	1
SMOOTHPRED	5	6	3	5	5
T3S	11	5	7	6	6
TYPEPARAM	2	3	1	4	3

Table 88: High accuracy ranking of each strategy for easy problems.

strategy	sharp peak	strong boundary layer	strong oscillatory	steep wave front	asymmetric wave front
ALTERNATE	9	2	12	10	11
APRIORI	11	13	1	11	10
COEFDECAY	5	10	9	5	7
COEFROOT	7	11	10	9	9
H&PERREST	8	9	8	6	4
NEXT3P	4	4	7	4	3
NLP	12	8	13	12	13
PRIOR2P	6	12	11	7	6
REFSOLNEDGE	1	1	4	1	1
REFSOLNELEM	2	5	5	2	2
SMOOTHPRED	13	7	6	3	5
T3S	10	3	2	13	12
TYPEPARAM	3	6	3	8	8

Table 87: Low accuracy ranking of each strategy for hard problems.

strategy	sharp peak	strong boundary layer	strong oscillatory	steep wave front	asymmetric wave front
ALTERNATE	13	1	12	3	4
APRIORI	8	13	2	9	9
COEFDECAY	10	8	7	8	8
COEFROOT	9	11	8	10	10
H&PERREST	6	9	10	7	6
NEXT3P	7	4	9	11	11
NLP	5	12	13	13	13
PRIOR2P	11	10	11	12	12
REFSOLNEDGE	2	2	3	1	1
REFSOLNELEM	1	3	4	2	2
SMOOTHPRED	12	5	6	6	7
T3S	4	7	1	5	5
TYPEPARAM	3	6	5	4	3

Table 89: High accuracy ranking of each strategy for hard problems.

strategy	nearly straight reentrant corner	wide angle reentrant corner	L-shaped domain	narrow angle reentrant corner	slit domain	mode 1 linear elasticity	mode 2 linear elasticity	battery	singular well	intersecting interfaces	multiple difficulties
ALTERNATE	11	11	11	12	12	12	11	10	12	12	5
APRIORI	10	9	5	3	3	2	8	5	6	2	10
COEFDECAY	3	1	1	2	2	3	3	4	5	3	8
COEFROOT	5	2	3	4	5	5	4	12	10	4	12
H&PERREST	1	3	7	8	8	8	5	1	3	11	2
NEXT3P	2	5	8	9	9	6	10	3	7	9	6
NLP	4	7	4	11	11	11	9	9	13	10	13
PRIOR2P	8	6	6	6	7	9	6	6	9	6	9
REFSOLNEDGE	6	4	2	1	1	1	1	11	1	1	1
REFSOLNELEM	9	8	9	5	6	4	7	2	8	8	3
SMOOTHPRED	12	12	12	10	10	10	12	8	4	7	7
T3S	13	13	13	13	13	13	13	7	2	13	11
TYPEPARAM	7	10	10	7	4	7	2	13	11	5	4

Table 90: Low accuracy ranking of each strategy for singular problems.

The ranking of the strategies indicates which strategies did best, but it does not indicate how much better one strategy is than another (or how close they are to being nearly the same). For this we can examine the factor by which N for a particular strategy is larger than N for the best strategy, as described at the beginning of Section 5. The factors are illustrated in Figures 609–614. Each circle represents the factor for one problem in the given category. If there is a number at the top of the graph, it indicates the number of factors that are larger than 10. The strategies that performed the best in that category have all the circles near the bottom of the graph, as in REFSOLN-EDGE, REFSOLNELEM and TYPEPARAM in Figure 609. To the right of the graph, the strategies are ranked according to the average of the factors for that category.

Based on the tables and figures in this section and Section 5.22, we make the following observations.

- REFSOLN-EDGE and REFSOLNELEM are the top two strategies in all categories except singular problems at low accuracy where they are in the top 5 with factors less than 2. Also note that REFSOLN-EDGE would have been the best strategy in that category if it had not performed poorly on the battery problem. The two strategies are equally good with each of them having the better average factor in three categories, and the largest ratio of their average factors being about 1.35. However, these strategies are considerably more expensive than most strategies.

- TYPEPARAM is the third best strategy in all categories of nonsingular problems, and is in the middle of the pack for singular problems where it has an average factor of 2.26 for low accuracy and 3.27 for

strategy	nearly straight reentrant corner	wide angle reentrant corner	L-shaped domain	narrow angle reentrant corner	slit domain	mode 1 linear elasticity	mode 2 linear elasticity	battery	singular well	intersecting interfaces	multiple difficulties
ALTERNATE	13	12	12	12	12	11	12	10	9	12	5
APRIORI	1	1	1	3	2	2	2	3	6	1	12
COEFDECAY	7	7	7	7	8	8	8	7	7	3	8
COEFROOT	4	6	6	6	7	10	6	12	10	6	9
H&PERREST	8	9	9	10	10	7	9	1	12	11	7
NEXT3P	10	10	11	11	9	13	13	8	8	9	10
NLP	11	11	10	9	11	6	5	4	13	10	13
PRIOR2P	6	4	4	5	6	9	4	9	11	8	11
REFSOLNEDGE	3	3	2	1	1	1	1	11	1	2	1
REFSOLNELEM	2	2	3	2	3	3	3	2	2	5	2
SMOOTHPRED	9	5	5	4	4	4	7	6	3	4	4
T3S	12	13	13	13	13	12	11	5	4	13	6
TYPEPARAM	5	8	8	8	5	5	10	13	5	7	3

Table 91: High accuracy ranking of each strategy for singular problems.

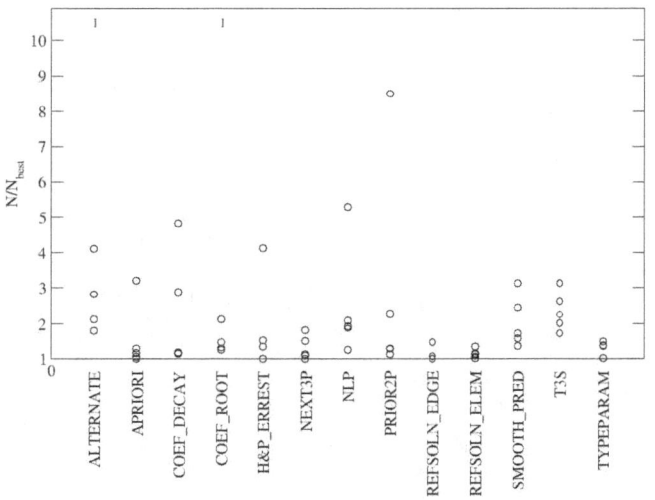

Figure 609: Factors by which N is larger than the best strategy for each easy problem at low accuracy. The table contains the average over all problems in the category.

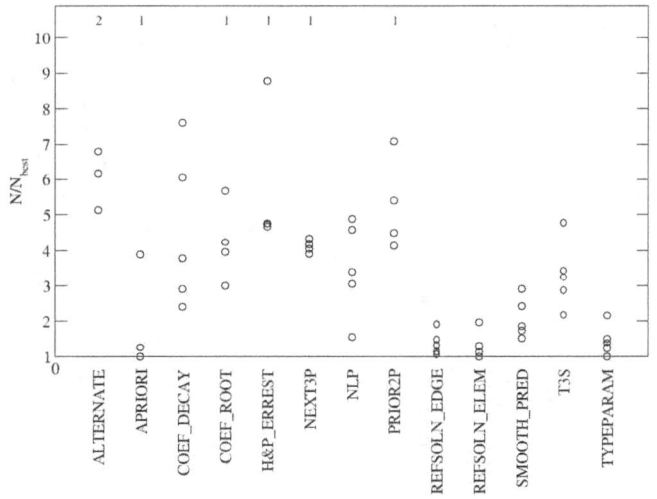

Figure 610: Factors by which N is larger than the best strategy for each easy problem at high accuracy. The table contains the average over all problems in the category.

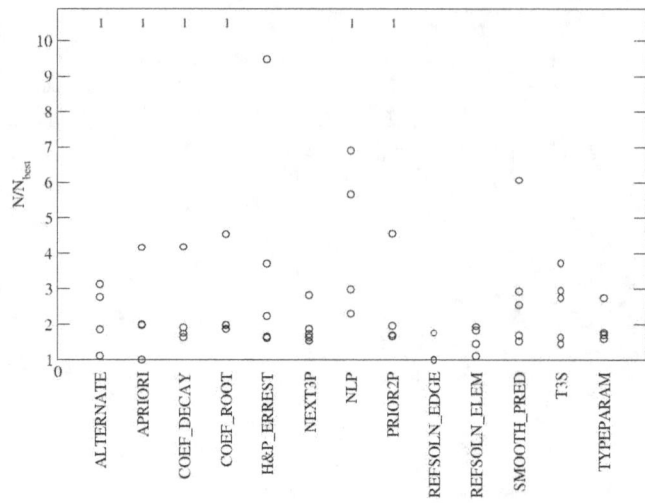

Figure 611: Factors by which N is larger than the best strategy for each hard problem at low accuracy. The table contains the average over all problems in the category.

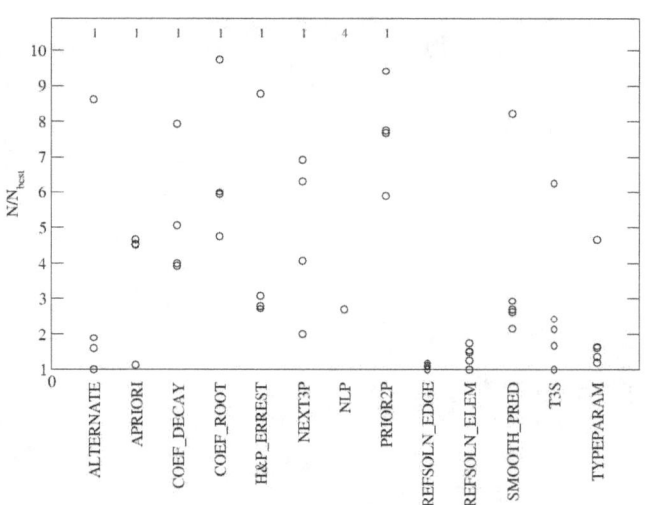

Figure 612: Factors by which N is larger than the best strategy for each hard problem at high accuracy. The table contains the average over all problems in the category.

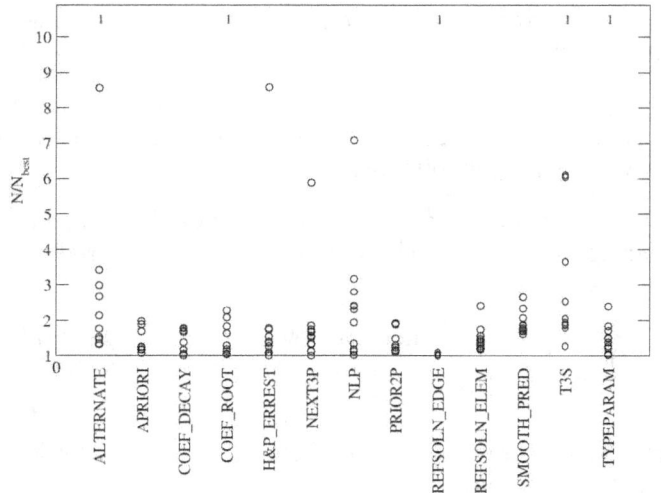

Figure 613: Factors by which N is larger than the best strategy for each singular problem at low accuracy. The table contains the average over all problems in the category.

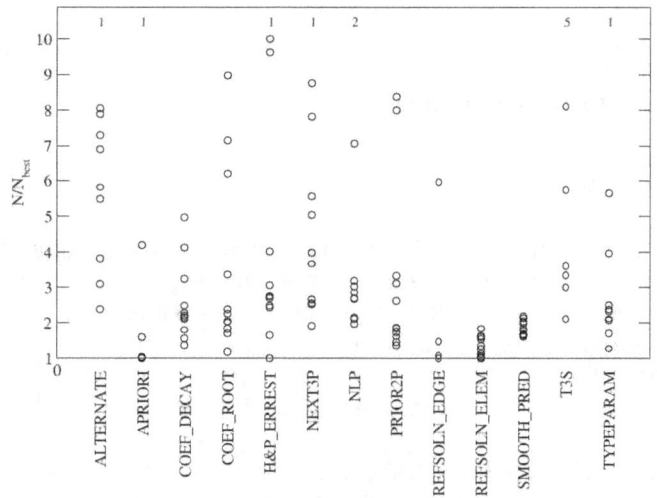

Figure 614: Factors by which N is larger than the best strategy for each singular problem at high accuracy. The table contains the average over all problems in the category.

high accuracy.

- SMOOTH-PRED is in the top 5 in all categories at high accuracy, and is the third best strategy for singular problems at high accuracy. But at low accuracy its average factors are in the middle of the pack and it is in the bottom four for many problems.

- APRIORI performs very well on singular problems with known point singularities and three of the five easy problems at both low and high accuracy. But it performs poorly on the hard problems, except for the strong oscillatory problem, and very poorly with the boundary layer.

- NEXT3P performs very well on nonsingular problems at low accuracy and fairly well on singular problems at low accuracy, but it is a bit more expensive than most strategies. It is below the middle of the pack at high accuracy with average factors around 5.

- T3S performs fairly well on nonsingular problems, but very poorly on singular problems where it has the largest average factor at both low and high accuracy, and is the worst strategy on about half of the singular problems.

- PRIOR2P performs poorly on nonsingular problems, but did very well on singular problems at low accuracy and fairly well on singular problems at high accuracy.

- COEF-DECAY is in the middle of the pack in all categories except for singular problems at low accuracy where it has the smallest average factor and is in the top four for most problems.

- H&P-ERREST is in the middle of the pack in all categories except easy problems at high accuracy where it performed poorly.

- COEF-ROOT performed poorly on nonsingular problems and is in the middle of the pack on singular problems.

- ALTERNATE performs very poorly in all categories, although it did well on a few of the hard problems.

- NLP performs poorly in most cases and is extremely expensive.

6 Conclusion and Future Work

In this paper we presented the results of a study of strategies for the hp-adaptive finite element method for 2D linear elliptic partial differential equations using newest node bisection of triangles. The hp-strategies are methods for determining how to select between the different possibilities of h- and p-refinement. Thirteen strategies were described and compared in a numerical experiment using 21 test problems. The primary metric for comparison was the convergence of the relative energy norm of the error vs. the number of degrees of freedom. A rough comparison of computation time was also presented, confirming the *apriori* expectations of the relative expense of the strategies.

We found that the REFSOLN-EDGE and REFSOLN-ELEM strategies performed best overall, in convergence, and are comparable to each other. However, they are considerably more expensive than other viable strategies. For problems with known point singularities and no other significant features, APRIORI appears to be the less expensive method of choice. For nonsingular problems, TYPEPARAM performs very well and is quite inexpensive. Another inexpensive strategy that performed very well at high accuracy is SMOOTH-PRED. Most of the other strategies have their good and bad moments.

Since the determination of what strategies to include in this study, other strategies have come to our attention or have come into existence. For future work we will extend the results of this study to include additional strategies as they are discovered. Also, we hope to use the lessons learned from this study to develop a better general purpose hp-strategy. For example, is it possible to get the excellent convergence performance of the reference solution strategies without the expense of computing the reference solution by combining some aspects of the reference solution strategies with some aspects of other strategies? Our conclusion is that, at this time, there is still much opportunity for the development of a general purpose hp-adaptive strategy that is both efficient and effective.

References

[1] S. Adjerid, M. Aiffa, and J. E. Flaherty, *Computational methods for singularly perturbed systems*, Singular Perturbation Concepts of Differential Equations (Providence) (J. Cronin and R. E. O'Malley, eds.), AMS, 1998.

[2] M. Ainsworth and J. T. Oden, *a posteriori error estimation in finite element analysis*, John Wiley & Sons, New York, 2000.

[3] M. Ainsworth and B. Senior, *An adaptive refinement strategy for h-p finite element computations*, Appl. Numer. Math. 26 (1997), no. 1-2, 165–178.

[4] ———, hp-*finite element procedures on non-uniform geometric meshes: adaptivity and constrained approximation*, Grid Generation and Adaptive Algorithms (New York) (M. W. Bern, J. E. Flaherty, and M. Luskin, eds.), vol. 113, IMA Volumes in Mathematics and its Applications, Springer-Verlag, 1999, pp. 1–28.

[5] R. Andreani, E. G. Birgin, J. M. Martnez, and M. L. Schuverdt, *On augmented Lagrangian methods with general lower-level constraints*, SIAM J. Optim. 18 (2007), 1286–1309.

[6] I. Babuška and M. Suri, *The* h-p *version of the finite element method with quasiuniform meshes*, RAIRO Modél. Math. Anal. Numér. 21 (1987), 199–238.

[7] ———, *The* p-*and* h-p *versions of the finite element method, an overview*, Comput. Methods Appl. Mech. Engrg. 80 (1990), 5–26.

[8] K. S. Bey, *An* hp *adaptive discontinuous Galerkin method for hyperbolic conservation laws*, Ph.D. thesis, University of Texas at Austin, Austin, TX, 1994.

[9] E. G. Birgin, *TANGO homepage*, http://www.ime.usp.br/~egbirgin/tango/.

[10] L. Demkowicz, *Computing with hp-adaptive finite elements, Volume 1, One and two dimensional elliptic and Maxwell problems*, Chapman & Hall/CRC, Boca Raton, FL, 2007.

[11] L. Demkowicz, W. Rachowicz, and Ph. Devloo, *A fully automatic hp-adaptivity*, J. Sci. Comput. 17 (2002), 127–155.

[12] T. Eibner and J. M. Melenk, *An adaptive strategy for* hp-*FEM based on testing for analyticity*, Comput. Mech. 39 (2007), no. 5, 575–595.

[13] W. Gui and I. Babuška, *The h, p and h-p versions of the finite element method in 1 dimension. Part 3: The adaptive h-p version*, Numer. Math. 49 (1986), 659–683.

[14] B. Guo and I. Babuška, *The h-p version of the finite element method. Part 1: The basic approximation results*, Comput. Mech. 1 (1986), 21–41.

[15] P. Houston, B. Senior, and E. Süli, *Sobolev regularity estimation for hp-adaptive finite element methods*, Numerical Mathematics and Advanced Appplications (Berlin) (F. Brezzi, A Buffa, S. Corsaro, and A. Murli, eds.), Springer-Verlag, 2003, pp. 619–644.

[16] C. Mavriplis, *Adaptive mesh strategies for the spectral element method*, Comput. Methods Appl. Mech. Engrg. 116 (1994), 77–86.

[17] J. M. Melenk and B. I. Wohlmuth, *On residual-based a-posteriori error estimation in hp-FEM*, Adv. Comput. Math. 15 (2001), 311–331.

[18] W. F. Mitchell, *PHAML homepage*, http://math.nist.gov/phaml.

[19] ———, *A comparison of adaptive refinement techniques for elliptic problems*, ACM Trans. Math. Software 15 (1989), 326–347.

[20] ———, *Adaptive refinement for arbitrary finite element spaces with hierarchical bases*, J. Comput. Appl. Math. 36 (1991), 65–78.

[21] ———, *A collection of 2D elliptic problems for testing adaptive algorithms*, NISTIR 7668, National Institute of Standards and Technology, 2010.

[22] A. A. Novotny, J. T. Pereira, E. A. Fancello, and C. S. de Barcellos, *A fast hp adaptive finite element mesh design for 2D elliptic boundary value problems*, Comput. Methods Appl. Mech. Engrg. 190 (2000), 133–148.

[23] J. T. Oden and A. Patra, *A parallel adaptive strategy for hp finite element computations*, Comput. Methods Appl. Mech. Engrg. 121 (1995), 449–470.

[24] J. T. Oden, A. Patra, and Y. Feng, *An hp adaptive strategy*, Adaptive Multilevel and Hierarchical Computational Strategies (A. K. Noor, ed.), vol. 157, ASME Publication, 1992, pp. 23–46.

[25] A. Patra, *private communication*.

[26] A. Patra and A. Gupta, *A systematic strategy for simultaneous adaptive hp finite element mesh modification using nonlinear programming*, Comput. Methods Appl. Mech. Engrg. 190 (2001), 3797–3818.

[27] W. Rachowicz, J. T. Oden, and L. Demkowicz, *Toward a universal h-p adaptive finite element strategy, Part 3. Design of h-p meshes*, Comput. Methods Appl. Mech. Engrg. 77 (1989), 181–212.

[28] A. Schmidt and K. G. Siebert, *a posteriori estimators for the h − p version of the finite element method in 1D*, Appl. Numer. Math. 35 (2000), 43–66.

[29] P. Šolín, *private communication*.

[30] P. Šolín, J. Červen´y, and I. Doležel, *Arbitrary-level hanging nodes and automatic adaptivity in the hp-FEM*, Math. Comput. Simulation 77 (2008), 117–132.

[31] P. Šolín, K. Segeth, and I. Doležel, *Higher-order finite element methods*, Chapman & Hall/CRC, New York, 2004.

[32] E. Süli, P. Houston, and Ch. Schwab, *hp-finite element methods for hyperbolic problems*, The Mathematics of Finite Elements and Applications X. MAFELAP (J.R. Whiteman, ed.), Elsevier, 2000, pp. 143–162.

[33] B. Szabo and I. Babuška, *Finite element analysis*, John Wiley and Sons, New York, 1991.

[34] R. Verfürth, *A review of a posteriori error estimation and adaptive mesh-refinement techniques*, Wiley Teubner, Chichester Stuttgart, 1996.

www.ingramcontent.com/pod-product-compliance
Lightning Source LLC
Chambersburg PA
CBHW081723170526
45167CB00009B/3676